THE UFO MAGAZINE UFO ENCYCLOPEDIA

UFO Magazine
UFO
Encyclopedia

Editor
William J. Birnes

Production
Nancy Hayfield

Associate Editors
Vicki Ecker • Don Ecker • Harold Burt • Ron Press

Contributing Editors
**Walt Andrus • Steve Bassett • Richard Boylan • Sean Casteel
C. Ronald Garner • William Gazecki • William Hamilton
Jolene Rae Harrington • Roger Leir • Melinda Leslie • Lex Lonehood
Philip Mantle • Mike Miley • John Schuessler • Karl Wolfe**

Pocket Books
New York London Toronto Sydney Tokyo

PHOTO CREDITS

All photos of Philip J. Corso courtesy of the Corso family. Unless otherwise indicated, all photos courtesy of UFO Magazine, Inc., copyright 2003. All rights reserved.

POCKET BOOKS, a division of Simon & Schuster, Inc.
1230 Avenue of the Americas, New York, NY 10020

For information address Pocket Books, 1230 Avenue of the Americas, New York, NY 10020

ISBN: 0-7434-6674-8

First Pocket Books trade paperback edition January 2004

10 9 8 7 6 5 4 3 2 1

Manufactured in the United States of America

Portions of this book were previously published in UFO Magazine.

A SHADOW LAWN PRESS BOOK

For Roy, August, and Casey

Your world, if you can take it.

Philip J. Corso

1915 • 1998

Acknowledgments

Over the eighteen-year history of *UFO Magazine* there are so many people whose work, support, encouragement, and inspiration I would like to acknowledge, it would take another volume of the *UFO Encyclopedia* to include all their names. Having said that, I would like to acknowledge the editorial staff and management of *UFO Magazine:* Vicki Ecker, editor-in-chief and founder; Don Ecker, news director and head of research; Sherie Stark, our first art director and founder; Ron Press, managing editor; Sharon Higgins, former managing editor, Katie Kuyper, editorial assistant and office manager; Don Silecchio, past president of UFO, Inc.; former production managers Bix Bigler and Kevin Convertito, former art director Kathy Allard; Rich Convertito, former head of circulation; Jay Eisenberg, advertising director; and current head of production and art director Nancy Hayfield.

I want to extend personal thanks to our researchers and writers over the years whose material, in one form or another, appears in this *Encyclopedia* or with whom I've had the pleasure to work. In particular, I acknowledge Chip Beck, Richard Hall, George Earley, George Straight, Michael Miley, Peter Robbins, John Greenewald, Jr., Peter Gersten, John Schuessler, Harold Burt, George Knapp, Melinda Leslie, Paul Davids, Bill Hamilton, Camille James, Stan Friedman, Kevin Randle, C. Ronald Garner, Dwight Schultz, Glen Boyd, Dan Fried, Jan Harzan, Bruce Maccabee, Roger Leir, William Gazecki, Jolene Ray Harrington, Linda Moulton Howe, Tim Iahn, Sean Casteel, Greg Bishop, Whitley Strieber, Phillipe Mora, Derrel Sims, Steve Bassett, Walter Haut, Glenn Dennis, Betty Hill, and The International UFO Museum in Roswell.

No acknowledgments list would be complete without a thank-you for keeping all of us in the UFO community honest to our friends at *Saucer Smear,* CSICOP, and *Skeptical Inquirer.* In particular, acknowledgments need to be handed to Michael Shermer, James W. Moseley, Karl Pflock, and, of course, Phil Klass. It is my fondest wish that Phil Klass will someday soon see the light and decide to 'fess up the the reality of UFOs and extraterrestrial contact. If he does, no doubt he'll want to write up his revelations in a blockbuster memoir, perhaps one that takes readers inside the CIA professional debunking establishment. If he does, he will probably need a literary agent, and he should feel free to call me.

My thanks and acknowledgments also go out to friends Bob Lazar, René Barnett, Bruce Burgess, Dr. Bob and Zoh Hieronimus, radio talk-show pioneer in our community Art Bell, and my friend George Noory.

My relationship with *UFO Magazine* began as a result of working with Lt. Col. Philip J. Corso on his book *The Day After Roswell.* Accordingly, his accomplishments and memory need to be acknowledged along with the continuing support of his son, Philip J. Corso, Jr. and his advisor and partner William Kent.

At Pocket Books, I would like to acknowledge my editor Mitchell Ivers and editorial assistant Joshua Martino and thank them for their patience and support during the compilation, writing, and production of this title.

Finally, I would like to thank my wife, Nancy Hayfield Birnes, art director and production manager both of *UFO Magazine* and this volume and my partner in Shadow Lawn Press and Filament Publishing, not only for her work and support in all of these efforts and for scores of other books, but, it can now be told, were it not for whose editing, review, and counsel, to me as well as to Phil Corso, there would have been no *Day After Roswell.*

William J. Birnes

FOREWORD

For over sixteen years, *UFO Magazine* has published stories about sightings of unidentified flying objects, flying triangles, alien abductions, time travel, exotic science, and government disinformation. It has carried on an honorable tradition of reporting on subjects the mainstream media wouldn't even touch no matter how reliable the witnesses. Or if the mainstream covered the story, it was with derision and sometimes even contempt. Within the past ten years, the media coverage of New Mexico Representative Steven Schiff, who asked the Air Force for the full story about Roswell, and Phoenix councilwoman Frances Barwood, who wanted full disclosure regarding the Phoenix Lights, painted both of these officials as somehow wacky because of their interest in UFO phenomena. But *UFO Magazine* honed to a straight line when it came to reporting news of the paranormal and continues to treat the stories in its genre as hands-off healthy journalistic inquiries.

The magazine was among the very first to cover the stories of the Phoenix Lights, the Kecksburg Incident, Men in Black, ongoing reports about cattle mutilations, and, of course the possibility of artificial structures on the moon as well as on the surface of Mars. *UFO Magazine* seeks to take a hard scientific approach to some of the most incredible reports and eye-witness accounts of our time.

The coverage of the Lonnie Zamora sighting outside Socorro, New Mexico; the Ed Walters Gulf Breeze Sightings in Florida; the Travis Walton "Fire in the Sky" abduction in Arizona, and John Keel's report of the Mothman sightings were all undertaken with a hard look at what facts there were, what facts could be ascertained from the first-hand reports, and what other factors might have influenced the way the stories reached the public. These first-hand accounts and other witness stories such as the abductions of Betty and Barney Hill, the extraction of strange metallic objects from under the skins of self-described alien abductees, and the horrific descriptions of mutilated cattle have fascinated readers and challenged skeptics to find explanations for these phenomena.

And then, of course, there was the magazine's coverage of the crop- circle story, the strange designs imprinted into standing grain at fields in England, Canada, and the United States. Some of the crop circles were obvious and admitted hoaxes, and readers will find entries here on the notorious Doug and Dave who bragged about their crop-circle-making exploits. But there are other crop circles that neither Doug nor Dave nor other hoaxers could have made. What might explain these? I know this is an ongoing story that fascinates my listeners as well as readers of the magazine.

Now, for the first time anywhere, a complete history of *UFO Magazine* and its most intriguing stories appear here in the *UFO Magazine UFO Encyclopedia,* a complete single-volume reference to UFO phenomena. What fascinates me the most about what you are about to read is not the fact that these stories have been compiled from years of research and witness interviews. What overwhelms me is the weight of the evidence, the consistency of the witness reports, and the rigid monotone of the official denials that something else is out there. In fact, I would go so far as to say that you cannot read this book and walk away without the nagging feeling that flying saucers are real, ETs might have already landed on the moon and fabricated structures there, our government and military have spent millions of dollars in psychic research, and that time travel is not only possible, we've been doing it for years.

Night after night on my show "Coast to Coast AM," I take phone calls from people whose lives have been touched by the paranormal. People have seen and been transfixed by huge, slow-moving black triangles floating noiselessly overhead. Indeed, even my predecessor Art Bell has reported an incident such as this as recently as the middle of November, 2003. Commercial and military pilots have reported seeing strange objects flying near them, but they have been told in no uncertain terms not to file official incident accounts. Even presidents Ronald Reagan and Jimmy Carter have made UFO sighting reports.

How can anyone, sifting through the amalgamation of this evidence, turn a blind eye or a deaf ear to what people have seen and said? I know I can't.

On my radio show I have described my own experience with remote viewing in an out-of-body experience I had when I was a teenager, an experience which I'm writing about in my forthcoming book, *A Worker in the Light.* On my show I've conducted experiments in focused consciousness to abate storms (such as Hurricane Isabel) that were relentlessly bearing down on populated areas, and to make the invisible Beltway Snipers appear visible to police. We've succeeded handily in both areas. Thus, I know that the paranormal is not just for bedtime stories on dark and stormy nights. It is as real as the chair you're sitting on.

Accordingly, I would like readers of this encyclopedia to hold their natural disbelief in suspension as they look through the hundreds of entries. Maybe the crash at Roswell is real, or maybe it's not. However, readers should ask themselves: Why would the wife of Chaves County sheriff George Wilcox, in her very own words written in 1947, describe what happened the night something crashed in the desert in terms of a flying saucer when there was no reason on earth for her to do so? Why would Army Air Force intelligence officers, men trained to distinguish between the most advanced and sophisticated aircraft, be unable to tell the difference between a weather balloon and an exotic craft? Why would an Army Air Force base commander, an officer who went on to become a four-star general, announce to the world that the Army recovered a flying saucer from the desert only to have the story revised the next day to become a recovery of a weather balloon? If this doesn't make sense to you, it doesn't to me either. If not a flying saucer, what was it?

This encyclopedia was edited by William Birnes, publisher of *UFO Magazine* and the *New York Times* bestselling author with the legendary Lt. Col. Philip J. Corso of *The Day After Roswell.* I've had Col. Corso's book on my "Nighthawk" radio show in St. Louis a number of times. As fantastic as the story see—the Army retrieved extraterrestrial technology from a crashed flying delta-shaped craft outside of Roswell in 1947 and reverse engineered this technology through American industry so as to develop the transistor, advanced microcomputer circuitry, laser cutting tools and weapons, and fiber optics—Col. Corso stood by it until the day he died in 1998. And whenever anyone managed to pull up Col. Corso's records, the man was where he said he was when he said he was. There's something about this story, perhaps the eerie convergence of the technology, that grabs you by the lapel and won't let you go.

When I first heard about this *UFO Magazine UFO Encyclopedia,* I was excited and waited with anticipation to read it. My wait was well rewarded. I only ask one thing: Suspend disbelief; read these pages with an open mind. And you can call me on "Coast to Coast" every night, but keep your mind and imagination flowing.

George Noory, Host
Coast to Coast AM

RAAF Captures Flying Sauce
On Ranch in Roswell Region

It all began . . .

Two enterprising young ladies met for lunch at a hamburger stand in West L.A. The era was mid-1980s, the season a cool, blustery spring. The two friends were Vicki Cooper and Sherie Stark, both bored veterans of a struggling business news bi-weekly called *Century City News.* In fact, Stark still worked there. Pulling up collars to avoid a chilly breeze, Cooper pitched her latest idea and Stark counter-pitched with value-added suggestions she felt she could offer. The following (abridged) conversation ran something like this:

"I'd like to start a magazine on UFOs," Cooper said, between bites of her burger.

"Great idea!" Stark said. "I want to help you."

Cooper then free-associated ideas that generally addressed the business side of things, which she openly—even proudly—admitted was not her area of expertise. Stark hinted that there may be room for their fledgling project at the offices of *Century City News.*

And that's exactly where infant *UFO Magazine* began to crawl, then walk. Cooper and Stark's first issues were pounded out on a Compugraphic Editwriter II, a hulking, two-component digital dinosaur that slowly spewed out acrid-smelling galleys to be waxed down on flimsy boards and photographed later at a nearby printer, transformed into what was to become the women's *raison d'etre* for years to come. But at the time, their familiarity with both the UFO field and publishing was slim-to-invisible.

Their first issue got a kick-start from local Billy Meier proponent Paul Shepard, who supplied stories, sources, and ads—and still owes the company money. It never occurred to Stark and Cooper that those exquisite, close-up photos of saucers allegedly photographed by a one-armed farmer in Switzerland (Meier) could easily be part of an elaborate disinformation gambit or well-funded hoax. The Billy Meier/Pleiadian UFO case was only one of many that were rife with propaganda both negative and positive, as well as some very suspicious characters. Only later would Cooper—always fighting timidity—nearly abandon the UFO ship because of the pernicious proliferation of spooks and psychos toxifying the entire field. But that's another story.

In the fall of 1986, Vol. 1 No. 1 was mailed out to four thousand UFO buffs ferreted from Cooper's small cache of saucer names and newsletters. Because nothing else like it existed anywhere—a magazine that dared to take a journalistic, non-tabloid approach to the subject—*California UFO* made a minor splash. (Cooper loved her native state and all that it implied, and so had attached it to the magazine's title.)

Publishing entrepreneurs Cooper and Stark later found out that by pinning the name of the greatest state in the nation to their magazine, national distribution was, at that time, out of the question. In short order, "California" was removed and the book became just *UFO Magazine.* In the interim, Cooper stumbled through the miasma of UFO—uh—stuff, selecting for publication what she thought might be a deft melding of hard information with the most popular, albeit "New Age" trends that, in the late '80s, so many of us were convinced foreshadowed an imminent and magnificent change of human consciousness, secretly being assisted by benevolent ETs and without a doubt ushering in the fullness of the Age of Aquarius! Stark tirelessly made calls and forged connections that would eventually make *UFO Magazine* a newsstand regular.

The magazine gingerly walked an editorial tightrope. Little did Cooper and Stark know that such things as dolphin intelligence, angel sightings, and space prophets named Uriel had little to do with the UFO phenomenon as generally regarded by those who know it best. So it was just about then that Cooper met Don Ecker, an ex-cop and former Green Beret who, on the surface, had little in common with a barefoot California chick in the throes of saucer mania.

Ecker's achingly direct, no-nonsense approach, though a tad abrasive at first, soon grew on Cooper and helped shore up what residual hard-nosed journalistic instincts she had left after swimming through gelatinous depths of UFO information, disinformation, and just plain crap.

Ecker's own encounter with a dying CIA agent, along with cattle mutilation investigations in Idaho, gave the former detective a solid grounding in the real facts of the UFO situation—and they weren't pleasant. The dark side of ufology was just then getting a significant boost, with the urgent musings of would-be spooks all but drowning out the more sanguine hype about space brothers and Strieberesque spirituality that tended to play up hopes and gloss over hard facts.

Arrogant boys heedlessly flirted with counterintelligence experts in hopes of cracking the mystery once and for all. The connection between intelligence deceptions and UFO matters became all too clear, and while Cooper folded herself into a metaphorical crouch, Ecker fearlessly interfaced with such rising "stars" as Bill Moore and John Lear and Bob Lazar. The result was *UFO Magazine*'s unrelenting coverage of the disinformation factor in ufology, but a steady decline in stories that relied largely on ingrained belief systems, whether religious, political, academic, or ufologic.

In 1990, Cooper and Ecker packed up and moved to the San Fernando Valley; by then they were tight friends and earnest lovers. But after West L.A., living in the Valley felt like being consigned to some biker-tinged boondocks—without good restaurants. Fortunately, Stark was still art director, and regular visits to her West Hollywood villa prevented the UFO couple from being totally choked off by Valley fumes. Rather unfortunately, the business side of *UFO Magazine* now came seriously under the purview of the right-brained.

At the same time, Don Ecker began hosting "UFOs Tonite!" on the Cable Radio Network, later called "Strange Daze" on Liberty-Works Radio Network. Both shows, though highly praised because of Ecker's solid broadcast appeal and careful selection of in-studio guests, were eventually cancelled.

It was touch-and-go for awhile, on the financial end. Though the magazine continued to be the best newsstand publication of its sort ever, some naive business decisions and the massive "laughter curtain," remorselessly preventing the subject from getting the mass attention it deserves, did keep the bottom line from being where it was supposed to be.

But *UFO Magazine* has always been blessed with a sort of synchronistic serendipity; in other words, the right people and the right things seem to come along at the right time. Thus, when Don and Vicki met Bill Birnes, new and more functional wheels of business were put in motion, greasing the coffers of the enterprise and bringing in fresh faces, fresh ideas, and even a fresh new office.

In its new incarnation, *UFO Magazine* happily gravitated back to home turf—West Los Angeles. This time, though, the office was allied with some other businesses, allowing cross-pollination of ideas that would serve to tighten up the magazine even more and introduce the beginnings of a truly world-class publication dealing with an amorphous topic.

In particular, the granddaddy of all UFO cases got an insider's scoop due to the new publisher, Bill Birnes. He co-wrote *The Day After Roswell* with the honorable Lt. Col. Philip Corso, covering never-before released aspects to the case that further heightened public interest in the government's UFO deceptions. "Whether you believed everything Corso said or not," says Birnes, "his story about reverse engineering technology that was 'not invented here,' backed up in his commanding officer's description of the program in his own memoirs, was the first time that someone with his military and government credentials and his high rank admitted that the United States government and military was involved with anything having to do with extraterrestrial origins."

In the last few years, *UFO Magazine* has managed to scoop all major media on topics of global concern: human cloning, Star Wars revival, and intelligent design, to name a few. One jump ahead of the mainstream, *UFO Magazine* maintains its position as the world's leading newsstand magazine on the subject.

We're not holding our breath, though. While many on the "UFO circuit" believe that final, open contact with the intelligence behind the phenomenon or official disclosure are surely imminent, seventeen years of reporting on the topic has convinced us that tactics used by military/intelligence, and the unwitting cooperation of many UFO buffs, may continue to keep the UFO matter just below mass perception and understanding. As such, the need for an objective publication about unidentified flying objects is more important than ever.

UFOs are real. For *UFO Magazine,* there is obviously no better *raison d'etre* than that.

Vicki Ecker, Editor in Chief
UFO Magazine

INTRODUCTION

Although *UFO Magazine* has published retrospectives and review issues of previous articles, never before have we attempted a full-blown encyclopedia, referencing not only our own material but material from other sources as well. This first edition of the *UFO Magazine UFO Encyclopedia* is our first attempt, but not our last.

Methodology

As we do in the magazine, we have tried to stay as neutral as possible regarding the credibility of different claims and assertions. Some of the entries will seem extraordinarily believable to readers, not only because the stories themselves have a ring of authenticity but because those who report the story or who have witnessed the event are credible themselves.

I've always been struck by the veracity of Officer Lonnie Zamora's report of his Socorro, New Mexico, encounter with a UFO and the entry that appeared in the private Ines Wilcox diary written in 1947 concerning the night Mac Brazel came to the Chaves County jail with a box of strange debris for her husband Sheriff George Wilcox's safekeeping. There are many other stories, not the least of which is the Cash-Landrum incident, in which the witnesses had absolutely no reason to falsify the truth and where the physical evidence supports their accounts.

Similarly, the story of abductees Barney and Betty Hill, who never sought the media limelight and who never even thought of telling their story in public, captured my imagination because it came at a time when mainstream America had never even heard of UFO abductions or the concept of "missing time." The Hills became public figures because of their abduction experience, and in spite of themselves. Thus, there is a kind of veracity to their story even though it sounds incredible.

On the other hand, readers will find some claims that, on the surface, seem impossible to believe. Will the world really end in 2012? Is there really a Majestic 12, half of whose members have voted to disclose the truth about UFO? Is our government actively engaging in dangerous time-travel experiments? Are there live creatures from humanity's own future being stored at Area 51, working with the government and using live human beings as guinea pigs in an incredible genetics experiment to fix human DNA gone awry in the future? Sounds like something out of a nightmarish and fiendish Governor Schwarzenegger movie, something so crazy—until it turns out to be true.

We have decided to be neither judge nor censor. Our limitations have not been credibility, but rather only space, because we want to present the broadest spectrum of claim, opinion, and anecdotal testimony to readers. Old hands in ufology may yawn at some of these stories because they've been told before. But for people wanting to know what's being claimed out there, we've adopted Associate Editor Harold Burt's approach that the total weight of evidence from hundreds of thousands of sources is just too compelling to disregard. Thus, we've been inclusive rather than exclusive.

In a court of law, only one or two eyewitnesses against the accused might be enough to substantiate the prosecution's charges. If that's the case, what about the hundreds of witnesses to a UFO event that the media and the government laugh off as a mass hallucination? What about the video of strange craft hovering over a populated area? These are the very events that took place over Washington D.C. in the 1950s, over Phoenix, Arizona, in 1997; over New York's Hudson Valley in the 1980s, and over Mexico City in the 1990s. As recently as late November 2003, during the lunar eclipse, witnesses along the entire eastern seaboard reported seeing a huge drifting craft. Thousands of witnesses attested to these events, perhaps enough witnesses to sway a judge and jury, and yet no one no one knows the real truth. Accordingly, we have gone with the witnesses whenever possible.

US Centric

There are UFO stories that originate from almost every country in the world. From Mongolia to Israel and from Brazil to Iran, witnesses have reported fantastic UFO sightings. UFO stories have been smuggled out of the People's Republic of China and appear in recently released KGB files from the Soviet Union. With a whole world's worth of UFO material, there is certainly a need for an international UFO encyclopedia. However, in this first edition, we have chosen to be US-centric simply because the bulk of stories from *UFO Magazine* have come from sources in the United States. In future editions, however, we will become more global in our scope and expand our focus to include material from Europe and the Third World.

Roswell

Both in the magazine and in this encyclopedia readers will find that we've spent a lot of time with the Roswell story. Indeed, the "Children of Roswell" entry is the longest entry in the book and comes right out of the "Children of Roswell" issue of *UFO Magazine.* To this day, the debate over what might actually have happened at Roswell, New Mexico in July, 1947 still rages, with some people saying a UFO crashed there and others saying it was only a top secret Army Air Force balloon.

One of the most perceptive comments about Roswell comes from Stanton Friedman, however, who has said on many occasion that in the early hours after the Army recovered whatever it recovered, they had no compunctions about calling it anything other than a flying saucer. There was no covering up of the facts, Friedman has maintained, because Col. William Blanchard told his Public Information Officer Lt. Walter Haut to release a story to the press that the Army had retrieved a crashed flying disk from the desert. In other words, until Army higher-ups stepped in and told Gen. Ramey to come up with a cover story, there was no attempt to hide the truth about Roswell. Thus, Roswell became one of the seminal events in American ufology, a nexus of UFO contact and government conspiracy to hide the truth. The story has become so rich and the witnesses so compelling, both in the magazine and here in the encyclopedia, we continue to focus on it in minute detail.

A Taxonomy of Ufology

Although we do not set forth a formal taxonomy of the field of ufology, we do include under the broad heading of ufology relationships among flying-saucer sightings, contact with extraterrestrial beings, alien and military abductions, out-of-body experiences and remote viewing, psychokinesis and extrasensory perception, contact with the world of spirits, and the long and ancient history of dream visions.

There are ufologists who focus specifically on sightings and who try to eliminate as many natural causes as possible. Their theory, and perhaps Bill Hamilton is as much a proponent of this as anyone, is that by rigorous examination of all relevant facts the UFO researcher should try to find a natural explanation for the event so as to eliminate it if at all possible. Only those sightings that can withstand such a rigorous tests deserve to be investigated further. Bill Hamilton brought this methodology to his research into the Phoenix Lights incident and that gave his research all the more credibility.

There are ufologists who focus on a history of records in national archives to see what light they can shed on the government's attempts, if any, to cover up whatever contacts it may have with UFOs or extraterrestrials. Among the foremost researchers in this field is Richard Dolan, whose work on UFOs and the national security state document the government's contortions as they try to come to some understanding of the UFO phenomena—even as they try desperately to keep what they know from the American people.

In the field of alien-abduction studies, investigations into the contacts between human beings and the creatures who abduct them, Budd Hopkins, Whitley Strieber, Dr. John Mack, and Professor David Jacobs are perhaps the best known researchers. Hopkins' books on the Copley Woods abductions, the Brooklyn Bridge abduction, and the phenomena of missing time have brought into the mainstream the claims of many who maintain they've been abducted by extraterrestrials. What Budd Hopkins has brought out into the light of day, novelist Whitley Strieber has popularized in his book, *Communion,* which reads like a gripping novel, but is, in reality, a recounting of his own experiences.

Dr. John Mack and Professor David Jacobs have conducted their own research with self-described abductees and found startling similarities among the different stories of people who could not possibly have communicated their private experiences to one another.

From the hard science of UFOs and the accounts of people who say they've experienced contact with them, ufology also embraces the spiritual side of the phenomenon, the links between human potentiality and a universe of aliens, spirits, demons, and sprites. This relationship between ufology and New Age is sometimes very troubling to the hard-science researchers, but is nevertheless a motivating factor for those who pursue alternative realities of the human psyche.

Finally, there is a political perspective to the UFO phenomenon, which we've tried to cover in depth both in the magazine and here. We've asked, if there is a real history of UFOs and their relationships with governments of the world, why would the government cover it up? There are lots of theories, some of which we've covered, that involve everything from private forces inside the government trying to corner the market on advanced alien technology to a general fear that if the world's populations knew for a fact that extraterrestrials existed all civil authority would break down.

Still another theory hypothesizes that fifty or so years ago, our government, and perhaps other governments, cut a deal with extraterrestrials to allow them to conduct experiments on human being in exchange for their agreement not to disclose their existence and at the same time advance our own science. A disclosure of this unholy alliance might so disturb the victims of that alliance that governments are loathe to reveal any information about what might have gone on. Nevertheless, lobbyists such as former candidate for the 8th C.D. Steven Bassett and Project Disclosure advocate Dr. Steven Greer argue that by assembling witnesses and presenting their stories in organized testimony to a congressional committee investigating the phenomenon, the government will eventually be forced to admit the truth about its dealings with UFOs.

The Truth

The truth about UFOs may well be that there is no truth until whatever extraterrestrials there are elect to disclose their own existence. If there is a government cover-up, the truth might lie in the branch of government that's conducting the cover-up. It might be that the cover-up is so diffused among different agencies that there can be no great revelation because what one agency decides to reveal other agency hides.

Maybe the Area 51 witnesses are correct and the experiments conducted at that secret Nevada installation are so potentially disturbing to humanity's sense of complacency that the truth can never be revealed. George Hoover, the United States naval officer who was given a copy of Morris K. Jessup's *The Case for UFOs,* and who subsequently annotated it with respect to theories about time travel, once said that the military had first-hand evidence of UFOs and ETs. To the military, the big secret wasn't UFOs themselves; however, it was the process of time travel and remote viewing which fascinated the military intelligence and research officers. Time travel, remote viewing, out-of-body projection, and psychokinesis, he said, became the collective holy grail of exotic science, which, if Area 51 witnesses can be believed, remains so today.

Perhaps this is the great truth that might someday be revealed.

William J. Birnes, Publisher
UFO Magazine

A-12 Oxcart

Prototype of the Lockheed SR-71 Blackbird, developed for the CIA as a high-altitude, high-speed strategic reconnaissance platform for spying on the Soviet Union. *See* **Central Intelligence Agency, History of UFOs in; Skunk Works; SR-71 Blackbird**

Abductees

Individuals who purport to have been taken aboard alien spacecraft against their will. Abductees often report being the subjects of medical experiments involving the extraction of body fluids and the use of genital probes, as well as having metallic devices implanted beneath their skin. Abductees have also told stories about witnessing or being subjects of fertilization experiments to hybridize both the human and alien species. Abductees have told therapists that they have seen their own hybridized children aboard spacecraft, and that their parents have been victims of abduction as well. This leads many abductees and researchers to believe that whoever is responsible for the abductions tends to track entire families through successive generations, suggesting that a gradual hybridization is taking place through each ensuing generation, either through transplanted DNA material or some other form of genetic manipulation. The apparent goal: to gradually blend humanity with extraterrestrial genetics.

Some abductee stories have become quite celebrated, such as those related by novelist Whitley Strieber about his encounters with an alien presence. A novelized version of his experiences and subsequent therapy to recover lost memories was the subject of his book *Communion* and the Philippe Mora film of the same name.

Abduction Scenario

Though unique to the individual, alien abductions seem to follow a prototypical scenario in which the individual, or small group of individuals, are isolated, paralyzed, and then, while completely conscious, physically transported aboard a spacecraft by some kind of light beam. The abductee often finds himself in the presence of the "typical" gray aliens. (These creatures sometimes appear to be insect-like as well; this is a common screen memory among abductees.) The creatures, although nonviolent, possess the extraordinary power of reducing the abductee's will to resist.

Actress Serena Roney-Dougal portrays Dr. Norma June Koster, a government pathologist, in the made-for-TV movie Arbiter Roswell.

Once aboard the craft, abductees are taken to medical examination rooms where they are secured to a gurney and probed through various body openings. The extraterrestrial examiners collect samples of sperm or ovum, skin and hair, and even fingernail shavings; they take measurements of their subject; they implant small devices beneath their subject's skin and often perform minor surgical procedures. If the subject has been previously impregnated by the aliens, as is often reported, the fetus may be extracted. The examination and erasing of the subject's memory is completed, and then the abductee is returned, usually to their original location.

In many cases abductees do not have conscious memories of the event, although subconscious images make their presence felt through dreams. Upon awakening from an induced sleep, subjects report feelings of anxiety—bordering on panic—that something terrible has happened to them. There are often physical manifestations of the experience, such as blood on pillows or bedclothes, a swelling underneath the skin, bruises that seem to have appeared out of nowhere, aches and pains, and, in the case of implantees, the psychological impression that there is a foreign presence within their bodies.

Because shame and humiliation are powerful emotions that can override rational thought, abductees feel violated and can become secretive and withdrawn, often exhibiting a variety of neuroses in their attempt to cover up the feelings they are experiencing. If the pain becomes too intense or the psychological dysfunctions too aberrant, abductees may seek some form of therapy. If the experiences they report under regression therapy defy reality, and most of them do, the therapist will seek a consult with another therapist better versed in the abduction phenomenon, and the experiencer will usually learn the truth about the events that have befallen him or her. Often at that point, the abductees will realize that they have been part of a long line of family abductions and will recover memories that go all the way back to childhood about strange visitors in the night.

Abduction Scenario, Physical Evidence

Abductees often point to physical evidence that confirms to them that their experiences have been real and not dream-induced. Some abductees report that they awake in the morning with pajamas or nightgowns that are turned backwards, as if they had been stripped naked during the experience and their alien abductors did not know how to put their clothing back on the right way. (This experience has also been reported by those who had been fully awake.) Other abductees have said that bandages covering their wounds had been removed and then replaced in a different location. Still others report observing a type of glow under their skins which they cannot wash off. A similar glow also appears in mirrors or on walls where the aliens have seemingly passed right through in order to get to the abductee.

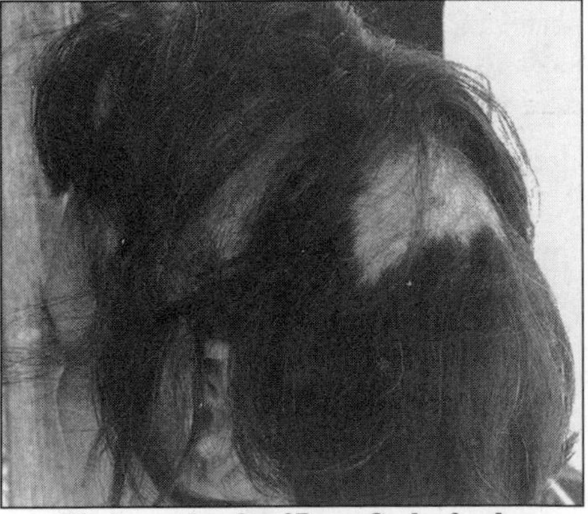

The burned scalp of Betty Cash after her encounter with a UFO in the skies over Dayton, Texas in 1980.

There's a great deal of skepticism regarding alien abductions, not only within the professional psychiatric community, but within the UFO community as well. The skepticism ranges from deep concern over the methodology utilized to recover memories from self-described abductees, to the range of experiences abductees report, to the lack of independent witnesses, and to the bizarre nature of the stories themselves.

Most memories of alien abductions are recovered via regression therapy: the reliving of events guided by a therapist. The therapist will either place the patient under hypnosis or administer a hypnotic drug. By and large, abduction experiences are invasive sexual-abuse experiences inflicted on vulnerable victims unable to defend themselves. Rather than face the realization that the abuser is a family member or a trusted family friend, the victim creates a screen memory to protect himself from the truth. That screen memory, prompted by years of viewing mostly hostile "aliens" on television or at the movies, becomes an alien who abducts the victim, deprives the victim of a will to resist, and performs sexual acts upon the victim before returning her to her own bedroom. However, many psychologists argue that this image is just that—an image—which is nothing more than a protective device the victim erects in his or her own mind. Unfortunately, because regression therapists often have their own agenda when beginning the therapy—that there truly is a hostile alien presence which abducts human victims—they impose this vision upon their vulnerable patients while they are in a particularly suggestible state. This can explain the commonalities noted among many abduction stories.

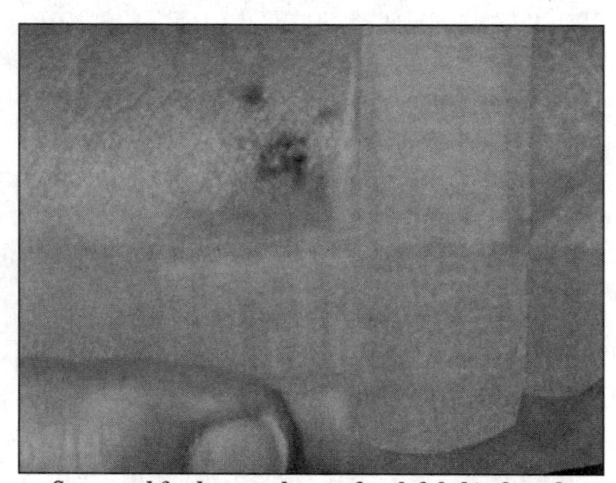

Scars and fresh wounds are often left behind on the body after an alien encounter. Abductee Pam Hamilton points to one such mysterious cluster on her arm.

The physical symptoms surrounding the abduction experience—the paralysis, the dream-like state, and the loss of awareness of the passing of time—all mirror those that people experience during periods of stress or overtiredness. For example, sleep paralysis is a common symptom when the person feels awake but can't move any part of his body. This is called a hypnogogic state, existing somewhere between sleep and waking. If this state precedes the dropping off into sleep, the person may interpret it as an outside force preventing him from moving. It is a common report made by abductees.

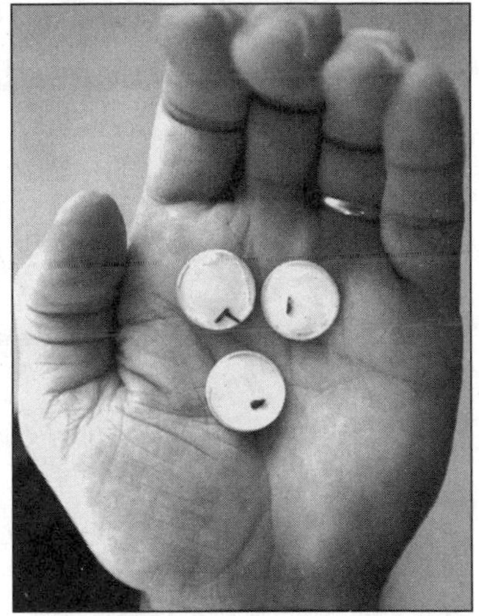

Alien artifacts recovered by Derrel Sims from the body of an alleged abductee.

While there are "traditional" abduction reports, there are also bizarre stories of people who say they've been abducted, indoctrinated by the space aliens into their cosmological philosophy, shown the future of planet Earth, and told to go back to report to others what they've seen. It's very easy to characterize these abductees as delusional because they often have no palpable evidence to support their stories. However, stories of prophets claiming otherworldly enlightenment and bringing back messages for the rest of humankind are common in all literatures, particularly the Bible, and have been the basis for all of the world's religions.

Among the scholars most noted for conducting abduction research and therapy are John Mack, David Jacobs, and Budd Hopkins.

Abductions, Indicators of

Compiled by abduction researcher and experiencer Melinda Leslie, this is a list of what victims, friends, or family should look for when they hear anecdotal stories of strange experiences from their loved ones. Some of the bizarre experiences that she red-flags as indicators of possible abduction cases are:

- *Missing or lost time, especially an hour or more.*
- *Unusual scars or marks, with no possible explanation of how the person received them. Indentations on the skin may look like small scoop marks, straight-line scars, scars on roof of mouth, in the nose, behind or in the ears, or in the genital area.*

- *Emotional reactions like fear, anxiety, or panic to the marks.*
- *Observing balls of light or flashes of light in the home or other locations.*
- *Memories of flying through the air that do not seem to be a dream.*
- *The person awakens with what psychologists call a "marker memory," an image that will not go away (such as an alien face, powerful overhead light, or unfamiliar or futuristic medical instruments).*
- *Observing beams of light which seem to extend through the walls into the person's room.*
- *Vivid dreams of alien-looking creatures or strange flying objects.*
- *The person has reported numerous UFO sightings, some of which may be accompanied by periods of missing time.*
- *Women who report multiple false pregnancies, or pregnancies in which the fetus simply disappears.*
- *People who awake in a place other than where they went to sleep, or don't remember going to sleep.*
- *Having dreams of especially prominent eyes, such as animal eyes (like an owl or a deer).*
- *Awakening startled in the middle of the night.*
- *Experiencing an aversion or strong attraction to pictures of aliens.*
- *Having phobias of heights, snakes, spiders, large insects, or being alone, or other issues related to personal security.*
- *Experiencing self-esteem problems along with any of the above phobias.*
- *Having blood or unusual staining on bedsheets or pillow, with no explanation as to how it got there.*
- *Compulsions to visit an out-of-the-way or unfamiliar place.*
- *Having the feeling of being watched all the time, especially at night.*
- *Hearing strange humming or pulsing sounds and being unable to identify the source.*

- *Awakening from sleep with repeated unexplainable nose-bleeds.*
- *Awakening from sleep with soreness in one's genitals that cannot be explained.*
- *Having back or neck problems (particularly T-3 vertebra), or awakening with an unusual stiffness in any part of the body.*
- *Suffering chronic and untreatable headaches, sinusitis, or nasal problems.*
- *Having electronics that malfunction without explanation.*
- *Seeing a hooded figure in your home, especially next to your bed.*
- *Frequent or sporadic ringing in the ears, especially in one ear.*
- *Suffering insomnia, or sleep disorders sometimes accompanied by traumas from terrifying dreams. This is also accompanied many times by a fear of falling asleep.*
- *Having paranormal or psychic experiences, or actually having these abilities.*
- *Being prone to compulsive or addictive behavior.*
- *Being afraid of what lurks in your closet, now, or as a child. The closet itself becomes a feared object.*
- *Fears of sexual relationships, or general relationship fears.*
- *Difficulties in trusting other people, especially authority figures.*
- *Suffering abnormal sensitivity towards certain lights or sounds.*
- *Having persistent or especially vivid dreams of destruction or catastrophe.*
- *Trying to resolve these types of problems with little success.*

Abramov, Alexander

A Russian scientist who discovered that there are large objects—obelisks—on the moon arranged in an artificial-looking grid pattern, with what seem to be measured distances between the objects, much like the pyramids in Egypt. These objects are not natural rock formations; rather they appear to be artifacts, the remains of structures built by an intelligent civilization, similar to the cliff dwellings in Mesa

Verde, Colorado. Abramov has applied the Egyptian pyramid pattern called an *abaka,* a grid of forty-nine squares, to the structures on the moon, and they match exactly.

Adair, David

An American scientist who testified before Congress in 1997 about recovered extraterrestrial spacecraft and how they are being reverse-engineered on earth. Adair's testimony included being taken to Area 51 in 1971 and shown an engine from a recovered alien spacecraft. He said that the engine had an organic sentience; it could literally feel and sense emotions and instructions from the pilot of the craft. There were organic-looking tubes surrounding the engine which mimicked the configuration of a brain stem and nerve fibers.

Though improbable-sounding, intelligence sources say the U.S. military currently has working sentient capability built into advanced, highly classified jet fighters, with engines capable of driving a spacecraft at light speed. *See* **Area 51**

Adams, William A.

Air Force Col. Adams was mentioned by Maj. Donald F. Keyhoe as one of the Air Force personnel invited by the CIA to view and analyze secret film of UFOs. Col. Adams was reportedly impressed enough with the film to have authorized continuing analysis of the material. Adams was also a participant in a series of classified meetings, the information from which was used in deliberations of the Robertson Panel. *See* **Robertson Panel**

Adamski, George

The first of the self-proclaimed UFO contactees. Adamski stated that he had been observing UFOs since 1946 and had his first encounter with extraterrestrials in 1953. He was a California figure of some prominence in the early 1950s, having reported over two hundred flying-saucer sightings around Mount Palomar. His notoriety grew, and he soon enjoyed a national following. In 1952 he claimed to have met humanoid aliens from the planet Venus and to have been taken aboard spacecraft and flown to other planets in our solar system, which he described in vivid detail in several books, including *Flying Saucers Have Landed.*

Adamski explained that he communicated telepathically and via sign language with one of the aliens. He was told that all aliens have

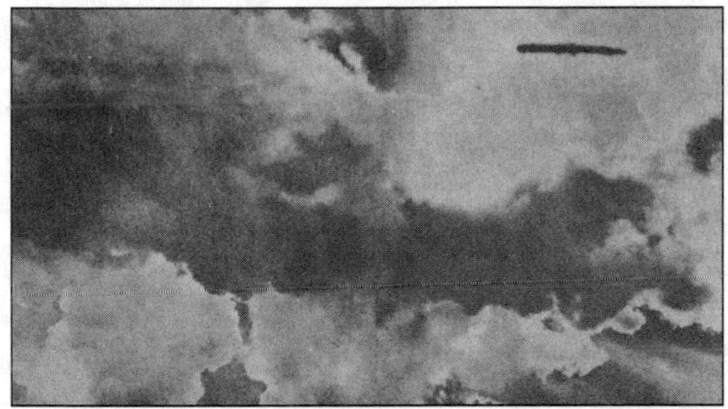

A Blue Book photo of an identified object from the early 1960s. Almost half of the sightings remain unexplained.

a human form, and that they had been taking earthlings away on space voyages for many years. Adamski claimed to have been taken on just such a trip, journeying to our moon and the planet Venus, where he said he met inhabitants of several different planets.

Controversy has always surrounded George Adamski and his claims. For example, detractors say that the conditions he described on the various planets are known to be impossible, while meanwhile, supporters respond that such absolute statements end up disintegrating in the light of new scientific discoveries, such as the recent evidence of microbial life on Mars. Such revealations have given Adamski's claims greater credibility over the years, and not less.

There were also disputes about the validity of photographs he claimed to have taken of the alien craft., yet U.S. and British intelligence agencies befriended Adamski and sponsored his worldwide speaking tour, leading to speculation that he was a government plant who, at a time when there were real UFO sightings that governments could not or would not explain, diverted attention from national security issues to tabloid adventures. To this day there are those who swear that everything Adamski said was true, and equally as many who say he was a con man. *See* **Goldin, Daniel**

Aerial Phenomena Research Organization (APRO)

Monitored "subversive" activities by the CIA in the 1950s. *See* **Central Intelligence Agency, History of UFOs in**

Aeroflot, Flight 2523

In one of the few witness-documented cases of UFO sightings coming out of the normally secretive former Soviet Union, the flight crew on this Aeroflot flight out of Voronezh on August 20, 1991, spotted a huge, bright saucer-shaped object shortly after takeoff. *See* **Voronezh, Russia**

Air Force Foreign Technology Division

Along with the Air Materiel Command, this was one of the first official branches of the military to study the UFO phenomenon. Former members of this division have said that the military had either retrieved alien remains and saucer debris or had made contact with alien spacecraft and extraterrestrials.

One former Foreign Technology Division officer, Lt. Col. Wendelle Stevens (Ret.) has stated: "I'm convinced beyond doubt that we have recovered aircraft, alien vehicles; that we have made contact with aliens, that we are communicating with them in some way or form, and that we have vehicles and bodies in preservation." Stevens has since been heavily criticized and characterized as a "hoaxer" by the community of UFO skeptics. *See* **Stevens, Wendelle**

Air Force, Studies of UFOs

For the last fifty years, U.S. Air Force has conducted extensive studies of the UFO phenomenon, although much of what they've allegedly found is still secret.

The cover of the infamous U.S. Air Force **Roswell Report;** *its official explanation for the Roswell Incident.*

Through a variety of operations—starting with the first analysis of UFO material by the Air Materiel Command under Lt. Gen. Nathan Twining in 1947, through Project Blue Book and into the present—the Air Force has actively studied UFOs while consistently denying their existence. The CIA's report on UFOs revealed that the public interest in these objects was used as a disinformation program to divert attention from test flights of the U-2 spy plane.

Maj. William T. Coleman, former Project Blue Book officer and Air Force spokesperson, quotes a 1962 Air Force order: "By this order, the Secretary of the A.F. of Information must delete all evidence of UFO reality and intelligent control, which would, of course, contradict the Air Force stand that UFOs do not exist. The same rule applies to A.F. press releases and UFO information given to Congress and the public." *See* **Central Intelligence Agency, History of UFOs in**

Air Materiel Command (AMC)

Under the command of Lt. Gen. Nathan Twining in 1947, the Air Materiel Command purportedly received debris from the crash of an extraterrestrial spacecraft at Roswell, New Mexico. Gen. Twining, in a memo whose veracity has since been challenged, wrote that the "(UFO) phenomenon reported is something real and not visionary or fictitious." The AMC made recommendations for the study of UFOs, since their extreme abilities to maneuver and take evasive action when spotted by terrestrial aircraft and radar lent credence to the belief that these objects were under the control of some form of intelligence.

Air Technical Intelligence Center

Based at Wright-Patterson Air Force Base outside of Dayton, Ohio, the Air Technical Intelligence Center was commanded by Lt. Gen. Nathan Twining in 1947. The debris from the Roswell craft was brought there for analysis in July of that year. *See* **Air Materiel Command; Twining, Nathan**

"A. K."

A photographer mentioned by nuclear physicist and UFO researcher Stanton Friedman, who said that in July 1947 he was stationed at Anacostia Naval Air Station in Washington, D.C., where:

> One morning they came in and they said, "Pack up your bags and we'll have the cameras there, ready for you." [They were then flown to Roswell Army Air Base.] We got in a staff car and we headed out, about an hour and a half, we was [*sic*] headin' north. We got out there and there was a helluva lot of people out there, in a closed tent. All kinds of brass runnin' around. And, they was tellin' us what to do; shoot this, shoot that! There was four bodies I could see when the flash went off. I remember they was thin and looked like they had too big a head.

See **Roswell, New Mexico**

Alamogordo National Laboratories

A government research center located near Roswell, New Mexico, where the first atomic bomb was developed during World War II. Alamogordo was the most top-secret research facility in the United

States in the early 1940s, and it was also the center of consistent UFO activity, especially in 1947. Alamogordo radar tracked unidentified objects overhead in July of 1947, one of which turned out to be the Roswell saucer that subsequently crashed. *See* **Roswell, New Mexico**

Albuquerque, New Mexico

Home of Sandia Atomic National Laboratories, an important government research facility, which was under the command of Gen. Robert M. Montague during the period of the Roswell crash. According to Stanton Friedman, Montague sat on the top-secret U.S. UFO command and control group, Majestic-12. *See* **Majestic**

Aldrin, Edwin "Buzz"

A member of the Apollo 11 crew, Aldrin was one of the first American astronauts to set foot on the moon with Neil Armstrong on July 20, 1969. There are some who say the entire Apollo lunar series was part of giant hoax to fool the world into thinking we actually landed on the moon. According to former NASA employee Otto Binder, the crew of Apollo 11 transmitted the following message to Apollo Mission Control as they orbited the moon:

Buzz Aldrin on the lunar surface. In Aldrin's face plate, some claim to see reflections of artificial structures: ruins of a former city or fortress on the moon.

> Oh, my God! You wouldn't believe it! These babies are huge, sir! Enormous! I'm telling you there are other spacecraft out there. Lined up on the far side of the crater edge! They're on the moon watching us.

This exchange was picked up by HAM radio operators using their own VHF equipment and did not go through NASA's own broadcasting network where, according to many UFO researchers and even military commentators, NASA routinely filters out information it does not want the public to hear.

During Apollo 11, according to some UFO researchers, when astronauts Armstrong and Aldrin planted the American flag on the lunar surface, it mysteriously flapped as if whipped by wind. This would be an impossibility because there's no atmosphere on the moon. Others have said that it was merely a spring snapping the flag out so that it would extend fully. But, according to still other critics of NASA, the reason the flags became spring-loaded after Apollo 11 was exactly so that gusts of wind on the moon *wouldn't* whip the flags. There were no springs in the flagstaff on the Apollo 11 mission. Other reports have said that Armstrong radioed Mission Control that UFOs were watching them as they walked on the moon.

It was in the reflection off the faceplate of Buzz Aldrin's space helmet, according to UFO researcher Richard Hoagland, that one can see actual ruins of a former civilization on the lunar surface, ruins which have been photographed by Russia's Lunar 9 and the United States Orbiter 2. *See* **Armstrong, Neil; Hoagland, Richard; Moon, Dark Side of the**

Alien Autopsy Film

Possibly one of the most elaborate commercial UFO hoaxes of all time, or a classic piece of deliberate government disinformation designed to throw the UFO community into confusion—or the real article. This film of a purported alien autopsy was supposedly presented to U.K. film producer Ray Santilli in 1995 by a former military cameraman known as Jack Barnett. Depicted in this film, consisting of ten reels of 16-mm black-and-white stock, is what is described as the autopsy of an alien corpse. The footage was presented as part of a documentary shown on the Fox Channel in the U.S. and scored high ratings for the network.

The grainy film, shot under dim lighting conditions with a single, often poorly focused camera, shows a group of apparent medical examiners wearing hazmat-type suits, performing an autopsy on a strange-looking creature said to be of extraterrestrial origin. Both skeptics and true believers have pointed to the same features in the film to prove their respective points of view. Skeptics argue that for a film of this import, the camera would most usually be placed on a stationary tripod to precipitate in-focus imaging,

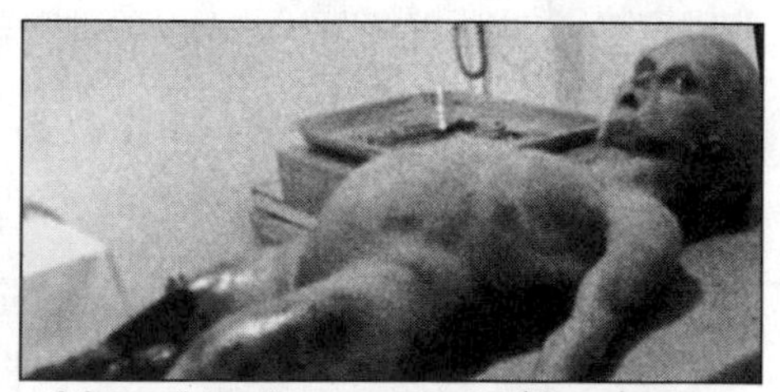

A frame from the controversial Alien Autopsy *film purporting to show a dead being from Roswell. Ray Santilli produced the footage, which was broadcast on the Fox TV network.*

and the footage would be shot using color stock.

They also raise questions about the "alien" undergoing examination—its characteristics don't entirely match those of other purported alien beings, especially the Roswell aliens. Still others have argued that there was no such person as "Jack Barnett" stationed where he claimed to be, and in addition, Santilli has reportedly changed his story about the film's origin a number of times. True believers argue that the hastiness of the autopsy required that the Army grab whoever was around to do the filming, and if they were trying to keep this film a secret, they would not go through channels with a request for an official photographer.

According to those who have researched the film extensively, Santilli claimed to have purchased the film in Cleveland, Ohio, in the summer of 1992 while researching a music documentary. The veteran cameraman from whom Santilli allegedly bought the film was said to have been a U.S. military cameraman who had been flown to Roswell, New Mexico, to film the crash, recovery, and autopsies of a UFO and its alien occupants. As part of Santilli's cash deal for the film, he was to keep the cameraman's identity a secret.

The film itself has been scrutinized by many experts. Some have pronounced it a fraud, others say that it is authentic. However, the one aspect of this case that puzzles most people is the reluctance of Santilli to have the film stock tested for age. This one fact alone has cast the most doubt on Santilli and his film.

Alien Beings

The UFO community believes there are many different species of aliens visiting earth, all of whom interact with human beings. Among the different races of aliens are small grays, large grays, reptilians, and Nordics. There have also been reports of other types of aliens—some appear as clouds of gas or have wolf-like faces, while others ply the galaxy and call themselves *Traders.* Swiss farmer Billy Meier, in one of the most controversial stories in UFO lore, claimed to have contacted a group called the *Pleiadians,* whose prophecies about humanity's future, some say, have largely come true. Other contactees, including Betty Hill, report having encountered aliens from Zeta Reticuli. Some of these alien races are purely benevolent, contactees say, while others, according to researchers like David Jacobs, are ultimately hostile. *See* **Extraterrestrials; Grays; Reptilians; Traders; Zetas**

Alien/Hybrid Children

Some UFO researchers have speculated that one of the ultimate goals of the alien races visiting earth is to hybridize their species with our own. Walt Andrus, retired director of MUFON, believes that this hybridization has been going on for successive generations in order to create groups of earth-linked aliens who will eventually take over the planet.

An alien-in-a-jar display from the UFO Encounters exhibition at the Los Angeles County Fair.

Abductees have described medical experiments in which their sperm or ova have been extracted for fertilization purposes, and they have then been re-abducted so the aliens could study the children who resulted from this breeding program. In fact, many abductees have been told that the tiny hybrid creatures the aliens have shown them are their own children. In one fascinating story, former Soviet pilot Col. Marina Lavrentevna Popovich has said that she has personally seen photographs of alien/human hybrid children. *See* **Abduction Scenario; Andrus, Walt; Jacobs, David; Lear, John**

Allagash Affair

In 1976, four Massachusetts friends: Chuck Rak, Charlie Foltz, and brothers Jack and Jim Weiner decided to take a camping trip. They left Boston and drove to a remote lake on the Allagash Waterway, where they set up camp on the beach. After building a large bonfire that they knew from experience would last between four and five hours, they climbed into two canoes and paddled out into deep water to do some night fishing. It was pitch black out on the lake and they planned on using the bonfire as a beacon to find their way back to the beach campsite.

Shortly after reaching the center of the lake, Chuck saw a huge ball of light hovering about two hundred feet above them, and he called the others' attention to it. They all then observed a bright, perfectly round sphere of light pulsating and changing colors. Charlie grabbed a flashlight and blinked it on and off at the object. The sphere, which had begun to rise slowly, immediately stopped and headed straight for the quartet. The men panicked and began paddling for their lives. As they attempted to flee, a cone of pale blue light shot out of the sphere and began advancing along the water towards their canoes. At that point they all blanked out.

The next thing any of them remembered was being back on the beach with the object hovering overhead. Standing there, dazed, they looked at the object for two or three minutes before it shot off at incredible speed into the night sky and disappeared. All four men felt strange and disoriented. Surprisingly, none of them felt like talking about their experience. They did, however, notice that their bonfire had burned down to only a few hot embers, despite their feeling that they had only been gone for about half an hour. This loss of sensation of the passage of time is a classic case of "missing time" in abduction scenarios.

Several years later Jim Weiner suffered a head injury that eventually led to temporo-limbic epilepsy, a syndrome in which the victim lapses into a dreamlike state even during his waking moments. Jim complained to his doctor about having very bizarre dreams and frequently awaking to find "strange creatures" in his room. Jim's doctor referred him to Raymond Fowler, a UFO researcher who had also studied the Betty Andreasson case. Fowler hypnotically regressed Jim, and the whole story about his camping trip to the Allagash Waterway came out.

Jim said that he and his brother and friends had been "picked up" on the lake by the beam of light, and that they were all taken aboard a spacecraft and examined. In similar hypnotic sessions, all four men confirmed the experience of the others.

They were able to provide detailed drawings of what they had seen, including the alien beings, their surroundings, and the instrumentation used for their examinations. It turned out that the aliens were particularly interested in Jim and his twin Jack, who, it seemed, had been abducted repeatedly almost since birth.

In 1988, Jack was abducted again along with his wife Mary. Shortly thereafter he found a strange lump on his leg which the family physician wanted to remove. Surgeons discovered a small, strange-looking object embedded beneath Jack's skin and sent the specimen to the Centers for Disease Control in Atlanta.

Later, Jack discovered that the object had been sent on to a military pathologist in Washington, D.C., and from there, an unidentified Air Force colonel had taken control of it. The object was never seen again. The full account of the abduction and its aftermath is recounted in Raymond Fowler's book *The Allagash Abductions*. *See* **Abduction Scenarios**

All-Union UFO Center

A UFO research facility and library in Russia which, among its large collection of books and papers on extraterrestrial phenomena, houses the work of Dr. Felix Zigel, who has documented over fifty thousand UFO sightings in the former Soviet Union, including declassified Soviet files which reveal that the Russians believed that their Mir space station was surveilled by UFOs. *See* **Russian Roswell; Zigel, Felix**

AMC

See **Air Materiel Command**

American Broadcasting Corporation Radio KSWS

The local radio affiliate of ABC in Roswell, New Mexico. KSWS was the first media outlet to learn of the Roswell crash on Monday, July 7, 1947. *See* **Roswell, New Mexico**

American Association for the Advancement of Science

In support of a UFO symposium at the second annual meeting of this organization in 1969, astronomer Carl Sagan, one of the major skeptical voices in the scientific community regarding the existence of UFOs, voiced support for a scientific study of those fields not associated with traditional science but of great public concern, such as "astrology, extrasensory perception, and unidentified flying objects, to show how these can be considered in a scientific way." *See* **Sagan, Carl**

Ancient Astronauts

An idea popularized by Erich von Däniken in his book *Chariots of the Gods,* which suggests that early extraterrestrial activity and manifestations might have been misinterpreted by our ancestors as visits from the gods.

His work has inspired such scholars as Zecharia Sitchin to examine the early origins of the human race. Published in 1968, von Däniken argued that spacecraft landed long before modern humanity peopled the Earth. The ancient visitors carried out breeding experiments and produced a creature intelligent enough to labor for their alien overlords. These new beings invented agriculture and became the first artists, then created their own warlike civilizations.

Von Däniken believes that images of ancient astronauts, complete with something resembling space helmets, were carved on stelae at the Mayan city of Copan. *See* **Pye, Lloyd; Sitchin, Zecharia; Däniken, Erich von**

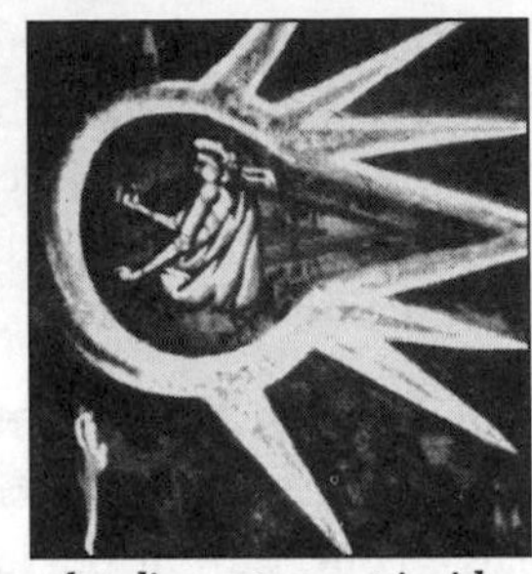

An alien astronaut inside a space capsule? Insert from a medieval mural hanging in Desani monastery, former Yugoslavia.

Andreasson, Betty

While living in Massachusetts, Betty Andreasson and her father observed strange-looking creatures wandering around outside their house, and she said she saw these creatures actually coming through the wall for her. In 1979, under hypnotic regression, Andreasson recalled being taken from her home by the "grays," brought aboard the creatures' spacecraft, and subjected to a variety of medical tests including the insertion of nasal and navel probes.

She recalled other images during her regression therapy such as being covered with fluid and breathing through a series of attached tubes, and then finally floating over a strange city Andreasson remembered looking "crystalline." She also reported that she heard the voice of God.

Although there has been considerable debate in the UFO community over the validity of Andreasson's claims, her story was similar to that of other abductees in many aspects. Her story was reported in Raymond Fowler's *The Andreasson Affair* and *The Andreasson Affair, Phase Two*, in which she told the author that she believed the visitations by aliens were benevolent and spiritual. *See* **Abduction Scenario; Fowler, Raymond**

Andrews, Colin

British researcher and author who is widely acknowledged as one of the world's experts on the crop-circle phenomenon. He is the founder of Circles Phenomenon Research International, the first organization established specifically to investigate the crop circle phenomenon.

His scientific investigations are responsible for much of the current information available on the subject and he was featured in William Gazecki's much heralded documentary, *Crop Circles, The Quest for Truth.*

British crop circle expert and researcher Colin Andrews.

Andrews is an electrical engineer by profession and a former senior officer in British regional Government. For three years he advised the British Government on crop circles, supplying technical and scientific reports to the undersecretary of state for the environment (Rt. Hon. Nicholas Ridely, M.P.) in the Margaret Thatcher cabinet. *See* **Crop Circles**

Andrews Air Force Base

Located outside of Washington, D.C., and one of the most important air-defense installations in the country since it protects the nation's capital, Andrews was the site of UFO sightings in June of 1952, when flying saucers flew over the Capitol and the Air Force was unable to intercept them.

Andrews radar control tracked the formation of flying saucers, as did other radar stations in the area, and personnel in the Andrews control tower made visual observations of these objects. The incident over Washington was widely reported at the time and is still the subject of debate between UFO believers and skeptics.

Andrus, Walt

Founder and international director of the Mutual UFO Network (MUFON), Andrus was its guiding force from 1969 through 2001, when he retired, leaving the organization under the leadership of current president John Schuessler. His endorsement of the validity of the photographs taken by witness Ed Walters helped lend credibility to the Gulf Breeze investigation.

Andrus observed a UFO over downtown Phoenix, Arizona, when he was a young adult, and that experience inspired him to conduct his investigation into UFOs and to start MUFON. Andrus believes extraterrestrials have long been on earth and have gradually occupied positions of economic and political power and plan to take over the planet.

Walt Andrus, founder and former international director of the Mutual UFO Network.

> I believe these alien creatures are hybridizing a species to create a generation of hybrids. With each succeeding generation of abductees, they are perfecting the hybrid species, recycling the DNA from abductee to the abductee's child and grandchild, until they have a race of human hybrids that will look no different from the way we do.
>
> However, they are still aliens endowed by an alien intelligence and whatever other powers the aliens possess. What is their mission? To insert themselves into powerful positions in government, in industry, in communications, and in the financial industry with each new generation. Theirs will be a "zero-force takeover." Therefore the threat may not be a threat of violence, but, rather, a loss of what we would consider our freedoms in a takeover of our planet.

See **Mutual UFO Network (MUFON)**

Anunnaki

The Anunnaki, according to scholar Zecharia Sitchin, were an extraterrestrial species that came to earth some 450,000 years ago and who genetically manipulated a hominid to short-circuit evolution and bring about homo sapiens. According to Sitchin, the Anunnaki's influence throughout human history can be seen in ancient world literature, particularly in Sumer, but also in such far-flung, astronomically aligned sacred sites as the Great Pyramids and Sphinx in Egypt, Stonehenge in England and its equivalent in the Golan Heights, and the Pyramids of the Sun and Moon at Teotihuacan. *See* **Sitchin, Zecharia**

Anza-Borrego Desert, California

For UFO and flying saucer observers, the Anza-Borrego desert in

Southern California is known as one of the hot spots for sightings. Many UFO enthusiasts set up cameras in this area to shoot footage of flying saucer activity in the skies over this desert. *See* **Mutual UFO Network (MUFON); National UFO Reporting Center**

Apollo Program

NASA's program of manned lunar missions, which, according to many UFO researchers, was canceled when it became clear that extraterrestrials had virtually "taken over" the moon by the time human beings had arrived. According to interviews with astronauts and reported records of Apollo-to-Mission Control radio transmissions, astronauts not only observed artificial structures on the lunar surface and alien spacecraft in orbit, but had even been fired upon by hostile aliens. During the Apollo 12 mission, for example, Donald B. Ratch, a UFO researcher who privately recorded this transmission, recalled the following exchange between the astronauts and Houston:

NASA's Voyager Mars space probe aeroshell photographed prior to its mid-1970s launch at Walker AFB (formerly Roswell Army Air Base).

> ***Apollo:*** *We have company.*
>
> ***Houston:*** *Say again?*
>
> ***Apollo:*** *I say we have company (short silence).*
>
> ***Houston:*** *You were told not to make transmissions such as that. Put it on the flight recorder and we'll discuss it when you get back (long period of silence).*

In addition to the alleged transmission between Apollo 12 astronauts and Mission Control regarding the presence of "company," there was another noteworthy incident. After Apollo 12's lunar module lifted off from the lunar surface and docked with the command module in orbit, it jettisoned the ascent stage, which then fell back

The official NASA crew portrait of Apollo 11: (left to right) Neil Armstrong, Michael Collins, Buzz Aldrin.

to the surface of the moon. Despite the relatively small size of the LEM, when it struck the surface, the entire moon reverberated for over eight minutes. According to some in the UFO community, this suggests the moon is hollow, and scientists at NASA have said, "It rang like a gong." If true, this phenomenon has not been fully explained to the American public.

In another theory about the history of Apollo, a spate of books and documentaries has attempted to convince the public that the Apollo astronauts *never* landed on the moon, and that NASA orchestrated the entire project from a movie soundstage, much like the 1970s television movie about a bogus Mars mission, *Capricorn One*. The pseudo-science and falsehoods perpetrated by the authors of this theory are laughable, but do highlight the extent to which the government might be willing to go to cover up the reality of the alien presence in outer space. *See* **Aldrin, Edwin "Buzz"**

APRO

See **Aerial Phenomena Research Organization**

Aquarius

See **Projects, Government UFO**

Area 51

The most famous top-secret military facility in the country, Area 51 lies adjacent to Nellis Air Force Base in Nevada, and is the location where the U.S. government supposedly has stored and maintained not only flying saucers but actual extraterrestrials.

Codenamed *Dreamland,* the facility, opened in 1955, is where test flights of the U-2, SR-71, F-117 Stealth fighter, and B-2 Stealth bomber first took place, but it is better known in UFO circles as the place where one could see flights of exotic-performing saucer-shaped craft. Downed or captured spacecraft are brought to Area 51 for analysis, according to UFO researchers, and it was this location where the controversial Robert Lazar said he worked on the reverse-engineering of alien technology until he breached security and was fired from the program.

Area 51 contains Groom Lake, which is six miles wide by ten miles long. The dry lakebed contains a constructed runway complex, the world's longest. The most secret part of the base, where the reported UFO activity takes place, is known as site S-4, which is said to contain hangars and laboratories to store and study alien saucers. The base is so important to American military aerospace programs that former president Bill Clinton exempted it from releasing environmental impact study reports to the public. Security at this base is so tight that signs posted at the perimeter warn would-be looky-loos that security personnel are authorized to use deadly force to prevent access to the base by intruders. *See* **Lazar, Robert**

Landscape of the Groom Lake Area 51 facility where, according to Robert Lazar and others, top-secret reverse engineering of alien technology is taking place.

Armstrong, Neil

The first human being to step onto the lunar surface during the Apollo 11 mission, astronaut Armstrong made the now famous quote: "This is one small step for man. One giant leap for mankind." Armstrong took the interesting photo of Buzz Aldrin in which strange spires seem to appear in the reflection of Aldrin's faceplate. Armstrong, in a story reported by two Russian space scientists who said he told it to them, also allegedly told Mission Control that two large space objects were watching the astronauts after landing near the lunar module. *See* **Aldrin, Edwin "Buzz"; Moon, Dark Side of the**

Arnold, Kenneth

A private pilot and member of the Idaho Search and Rescue Mercy Flyers, who, on June 24, 1947, joined a search mission to find a missing Marine transport plane that was believed to have crashed in the Cascade Mountains of Washington. While flying at nine thousand feet, Arnold saw nine silver disks moving in formation directly across his flight path. He later estimated their speed at over 1,700 miles per hour, a thousand miles per hour faster than the world airspeed record at the time.

Kenneth Arnold. His 1947 report of crescent-shaped UFOs over the Cascade Mountains in Washington inadvertently helped coin the term **flying saucer.**

Arnold was not the only witness to these craft. Pilots flying a nearby Douglas DC-4 also saw the disks, and they were seen by hundreds of people on the ground. In an interview with the *East Oregonian* newspaper, Arnold said: "They flew like a saucer would if you skipped it across the water."

When the Associated Press picked up the story, they called the disks "saucer-like objects," from which the term "flying saucer" officially became an integral part of our language.

Arnold, Steve

See **Kaufman, Frank**

Atomic Bomb

The development of the atomic (and hydrogen) bomb is one of the reasons, UFO researchers say, extraterrestrials have decided to intervene in our affairs. The destructive nature of these weapons is so immense that a full-scale nuclear war is more than capable of wiping out all life on the planet, an event, according to contactees and abductees, the aliens will prevent by direct intervention and disclosure, if necessary.

Atrevida

A Spanish warship from which crew members were able to observe a spectacular display of UFOs on June 22, 1976, over the Canary Islands. *See* **Canary Islands Case**

Aurora

Reportedly one of the U.S. Air Force's most top-secret aircraft, Aurora is said to be propelled by a pulse engine fueled by cryogenic (hyper-chilled) methane. Its contrail looks like donuts strung on a rope, and these have been observed over the skies of Southern California as well as at Area 51, the aircraft's supposed base. Aurora's estimated top speed is over six thousand miles per hour, and it uses advanced aerodynamics that some say were reverse-engineered from flying saucers. Aurora is supposedly very difficult to control, which has resulted in a number of crashes and fatalities. *See* **Area 51; Skunk Works; SR-71 Blackbird**

AVRO

Man-made flying saucers developed and manufactured in Canada during the 1950s. These were based, in part, on early German designs for disk-shaped craft that proved difficult to control and were never fully developed. Although they never became operational aircraft, Air Force Lt. Col. George Edwards said that the AVRO devices were used as "covers" for the test flights of real flying saucers the Air Force had acquired.

Ayres, William H.

A member of the House of Representatives from Ohio who revealed that "Congressional investigations have been held and are still being held on the problems of unidentified flying objects. Since most of the hearings are classified, the hearings are never printed."

Azhazha, Vladimir

A scientist, oceanographer, and former Soviet submarine captain, Dr. Azhazha stated: "UFOs transmorph, going from saucer shape to cigar shape to a spiral in minutes. They can materialize and dematerialize at will. The craft and occupants are varied and may be from dozens of different sources and civilizations."

B-2 Spirit Stealth Bomber

Resembling the crescent-shaped outline of the Roswell spacecraft, the B-2 Spirit Stealth bomber was developed and tested at Area 51 under tight security. Though untrained observers may have confused the B-2 with a UFO because of its odd shape, the bomber is essentially a flying wing resembling the designs of Jack Northrop before World War II and then tested after the war. *See* **Flying Triangles**

B-29 Bomber

When Maj. Jesse Marcel of the 509th Bomber Squadron, headquartered in Roswell, New Mexico, was ordered to deliver the Roswell spacecraft debris to Brig. Gen. Roger Ramey at Fort Worth Army Air Field in Texas, Marcel described the load of debris as "half a B-29 full." UFO historians maintain that if the Roswell debris was, in fact, a weather balloon, as the Air Force maintained up until 1996, it could not have filled up half a B-29.

Bacellar, Carlos Alberto

Captain of a Brazilian ship sailing in the South Atlantic on naval exercises, who witnessed multiple flying saucers overflying his vessel on January 6, 1958. This incident, witnessed by others as well, became known as the *Brazilian Navy Case. See* **Brazilian Navy Case**

Baikonur Cosmodrome

The primary Soviet/Russian space-launch facility. In June 1982, UFOs were sighted flying over the area as if they were observing activity at the cosmodrome. The installation reportedly suffered damage from the UFOs' energy fields, some Soviet officials later told the press, although it was never determined whether the UFOs were, in fact, responsible or whether the damage was preexisting.

Bakersfield, California

Site of an reported UFO sighting on June 14, 1947, just weeks before Roswell. Richard Rankin, a passenger aboard a Chicago–to–Los Angeles flight, observed a triangular formation of ten disks through the airplane window. Twenty-five years later California Governor and later U.S. President Ronald Reagan also reported seeing a UFO "zig-zagging round" in the skies near Bakersfield. *See* **Reagan, Ronald**

Balducci, Corrado

The Vatican's expert on exorcism and research into the paranormal, who, on April 5, 1998, announced that, from the Holy See's perspective, contact with extraterrestrial "beings" was a current and ongoing process.

Speaking in a rare candid interview, Monsignor Balducci explained the Vatican's official position about the visitor phenomena to a television audience by saying, "It is very real." He then went on to reveal an extensive Vatican international intelligence organization, which, via its listening posts at their embassies around the world, had received extensive information about extraterrestrials and their contact with human beings. Balducci also revealed that he is a member of a Vatican council with responsibility for developing explanations for the existence of extraterrestrial beings within the context of Roman Catholic dogma. *See* **Vatican**

Ball Lightning

An atmospheric phenomenon which has often been confused with UFO sightings, ball lightning does not manifest the typical characteristics of lightning. Instead, it usually appears as a glowing sphere of plasma which drifts horizontally through the air and can seem to hover in one spot. This has led some observers to believe the balls are actually either sentient beings themselves or remotely piloted observation craft from other planets. Ball lightning is typically the size of a grapefruit, but can be as small as a pea or as large as a bus. The balls sometime seem to be suspended only a few feet or tens of feet in the air, but it can also bounce along the ground, which makes them appear to observers as if they're searching for something. In the vast majority of cases, the phenomenon lasts only a few seconds.

Balloon Bombs

Capable of crossing the Pacific Ocean, balloon bombs were launched by the Japanese against the U.S. during World War II. One theory suggests that the craft that crashed in Roswell in 1947 was one of these bombs, *aka* known as *fugo* balloons, which had somehow managed to stay aloft for two years after the war.

Bando

See **Projects, Government UFO**

Barauna, Almiro

A civilian whose photographs of the UFOs in the Brazilian Navy Case confirmed, at least to some of the witnesses, that the objects they had observed from the deck of their ship were real. *See* **Brazilian Navy Case**

Barclay, John

A witness to the appearance of a strange-looking airship in April 1897 over the skies of Rockland, Texas, during what has come to be known as the *Great Airship Mystery* of the late nineteenth century. Barclay reported that the object made a strange noise. He then observed a long, blimp-shaped object hovering overhead moments before it landed. A man, whom Barclay believed had come from the craft, appeared and paid Barclay to obtain some items for him from a local store. After Barclay returned with the items, the man said he was leaving for Europe, took off, and sped out of sight. Some researchers have speculated that the craft was more likely a time ship from earth's future than a spaceship. *See* **Great Airship Mystery**

Barnes, Harry

A radar controller at Andrews Air Force Base on duty during the 1952 Washington, D.C., UFO sightings. Barnes reported blips on his radar screen which confirmed the existence of UFOs that witnesses were able to see with their naked eyes and that were also captured in photographs and news film footage. *See* **Andrews Air Force Base**

Barnett, Grady

A soil engineer for the U.S. Conservation Service who came across the San Augustin UFO crash site around July 3, 1947. San Augustin was, reportedly, the "other" crash site linked to the Roswell crash. When Barnett came upon the debris field, he found other people, including some archaeologists, already combing over the wreckage. No Army teams had yet arrived. Barnett said he saw a large metallic object partly buried in the ground, though mostly intact, with dead bodies scattered around outside the craft. The bodies were wearing silver, form-fitting jumpsuits. Shortly afterwards, a military team arrived, cordoned off the area, and warned Barnett never to discuss what he had seen. *See* **San Augustin, New Mexico**

Barwood, Frances

Frances Barwood

Dubbed by the press "Beam Me Up" Barwood because of her public outspokenness after the 1997 Phoenix Lights sightings, Phoenix City Councilwoman Barwood was subject to a recall vote to remove her from office. The Phoenix Lights incident involved the appearances of mysterious objects that hovered in the skies over Phoenix on successive nights during the spring and early summer of 1997. Barwood believed that her criticisms of the Mayor's office, then-Governor Fife Symington, and the military—all of whom seemed to ignore the existence of the lights even though they were broadcast on national television—were at the core of the recall vote.

Barwood survived the recall and ran for the office of Secretary of State on the platform of, among other things, the issue of government secrecy. She became known as the "UFO candidate," and her campaign attracted a lot of attention. UFO conspiracy theorists attribute her sudden death from cancer after the campaign as part of a larger government plot to silence all of those who speak out on the issue of UFOs. *See* **Phoenix Lights**

Bassett, Steve

Steve Bassett

An independent candidate for the Eighth Congressional District in Montgomery County, Maryland. In the November 2002 election, Bassett ran on a UFO-disclosure platform, the first candidate to run for Congress with a campaign demanding that the government reveal what it knows about the existence of UFOs and contact with extraterrestrials. Bassett has described himself as a UFO lobbyist intent on prying UFO secrets out of recalcitrant government officials.

Beamships

These flying saucers look like an inverted bell with the top of the dome squared off. Their most prominent feature is what looks like three equally spaced balls attached to the bottom of the saucer. *See* **Meier, Eduard "Billy"**

Bean, John

A pilot who witnessed a UFO during one of the most well-documented California sightings near Livermore in 1949. Bean was driving along the road when he stopped to retrieve something from his briefcase. At this point, while watching a commercial passenger airliner landing at a nearby airport, he observed a circular UFO, on a near-collision course with the airliner, execute an impossibly tight turn and accelerate straight upwards to avoid hitting the landing plane.

Beardsley, James

An FAA traffic controller, Beardsley was riding in the cockpit of a commercial flight when he observed a near-collision between his aircraft and a formation of four UFOs. Beardsley filed an official report with the FAA, which was not released to the public for a number of years. UFO researchers cite the credibility of Beardsley's observations, his position in the cockpit as a pilot-observer, the proximity of Beardsley's plane to the UFOs, and the confirmation of the presence of UFOs by ground radar as an example of the kind of observation that aerospace and aviation professionals have made over the years, which the government routinely keeps secret.

Beaune, France

On a medieval tapestry housed in the Collegiale Notre Dame in Beaune, there is a depiction of the Virgin Mary against a background within which there is a rendition of an apparent flying saucer. UFO researchers believe that this and other medieval and Renaissance works of art depicting clear images of flying craft with or without pilots aboard are clear indicators of the presence of either extraterrestrial spacecraft or time machines during the early history of civilization.

Belenko, Viktor

A Soviet MiG-25 pilot who defected to the United States via Hokkaido, Japan, where he landed his jet in 1976. Belenko later became a consultant to Air Force and Navy combat training institutions such as Top Gun and wrote about his January 1972 encounter with a UFO in an article in *UFO Magazine*. Belenko never reported the encounter to his Soviet flight controllers, since it would have guaranteed his never sitting in the cockpit of a MiG fighter again.

Belgian Triangles

The triangular-shaped craft sighted repeatedly over Belgium from November 1989 through April 1990. Not only did private citizens observe these huge objects, but NATO F-16 fighters were scrambled to intercept them. The craft were able to outfly the F-16s and speed off, no mean feat considering that fifteen years ago the F-16 was one of the fastest aircraft in the world. It was estimated that almost nine out of every ten people in Belgium saw the flying triangles during this period, and the overwhelming number of witnesses and witness reports for both day and night sightings posed such a challenge for the Belgian government that, rather than simply ignore the reports or deny the existence of the craft, Belgium became the first and only nation in the world whose government openly and officially recognize the existence of UFOs. *See* **De Brouwer, Wilfred; SOBEPS**

Bell, Art

Art Bell (and friend) during a live broadcast from his studio in Pahrump, Nevada.

Radio talk-show host Art Bell is a modern pioneer of alternative talk-back radio. Following in the tradition of Long John Nebel and Barry Farber, Bell transformed late-night radio from a soapbox for ranting paranoids into a place where anyone with a strange or bizarre encounter with the paranormal could find a friendly (though sometimes skeptical) ear and his or her night in court.

"The Art Bell Show" has helped to introduce the research of Richard Hoagland to a national audience, and it was also a venue where controversial guests such as Area 51 engineer Robert Lazar, remote-viewer Ed Dames, and alien abductee Whitley Strieber made their earliest appearances. After a years' absence for medical problems,

Bell returned to regular broadcasting in 2003 as the weekend host of George Noory's "Coast to Coast A.M." *See* **Noory, George**

Bender, Albert K.

The editor of *Space Review* magazine and the founder of the International Flying Saucer Bureau, who, in 1953, came into possession of what he called "vital information" proving that the U.S. government was covering up the existence of UFOs. Bender had written several articles which were scheduled to appear in a forthcoming edition of his magazine when three men dressed in black suits and sunglasses showed up at his door unannounced. They told him they had read his articles even though the stories had not yet appeared in print, and said that although his information was accurate, it was too sensitive. According to Bender, the "men in black" warned, "We advise those engaged in saucer work to please be very cautious." They scared Bender so badly he retired from UFO investigation. *See* **Men in Black**

Bentwaters Case

Two of the most noteworthy UFO encounter cases took place almost twenty-five years apart at the Royal Air Force Base at Lakenheath-Bentwaters, an important NATO airbase in England. The first set of encounters occurred on two successive nights in 1956. On the second night, August 13, at 10:55 P.M., base radar picked up unidentified targets traveling at 2,500 miles per hour overhead. Controllers in the Bentwaters tower saw bright lights flash by, as did a transport pilot in the immediate vicinity.

An object was then spotted hovering stationary in the air just outside the base. The object instantly accelerated to supersonic speeds, zigzagging across the airspace, then just as quickly stopped in mid-air. The Lakenheath watch commander scrambled a Venom night-fighter to investigate. For the next half-hour, the jet tried to intercept the bright object but was consistently outmaneuvered. The RAF pilot finally had to take evasive maneuvers to keep the object off his tail. The jet ran low on fuel and returned to base; a second Venom fighter was sent aloft to track the UFO, but it soon developed instrument problems and had to return to base as well. The UFO remained in the area for a short while longer and then disappeared.

The second and even more celebrated incident at Bentwaters took place on December 26 and 27, 1980, during the height of the Soviet invasion of Afghanistan. This second case would become England's most famous UFO encounter and the subject of several books, including Larry Warren and Peter Robbins' *Left at Eastgate* and Jenny Randles' *UFO Crash Landing?* which documented the events of the night American Air Force Lt. Col. Charles Halt supposedly had a face-to-face encounter with an extraterrestrial. Whether a real UFO/ET encounter, a mass hallucination, or a deliberate and elaborate psychological stress test of the vulnerability of Air Force personnel, complete with holographic projections and special effects—the entire RAF base reacted to a security breach which, authors Warren and Robbins claim, turned out to be the landing of an extraterrestrial spacecraft.

During the early morning hours of December 26, 1980, people who lived in the vicinity of the base reported seeing lighted objects fall out of the sky. These objects were tracked by an array of British and American military and civilian flight-control radar. The lights also attracted the attention of two U.S. Air Force Air Policemen on patrol near the Rendlesham Forest gate of the air base who believed it was a downed aircraft. Air Force personnel searched the forest by truck for debris but instead came upon a strange hovering light that seemed to react to the patrol by increasing the intensity of its glow the closer the patrol approached. The Air Police detail phoned in the report to the air base's security division, which quickly sent units to the scene.

Two members of the security detail approached the object on foot; one of them saw that it had a glass-like surface, which bore strange markings. The men could also see that, rather than hovering above the ground, the object was resting on three landing-gear struts. These were suddenly retracted, and the object then levitated a few feet above the forest floor and began to glide over the men's heads, weaving its way through the treetops as it gained altitude. In an instant, the object shot into the air and disappeared. The patrol returned to their headquarters to report the encounter.

The UFO's trajectory toward Rendlesham Forest had brought it dangerously close to the highly secure air base. The concern of base security was also heightened because radio contact with the security detail sent to investigate the glowing light was lost once their truck entered the forest, and one of the air base's jets that had been deployed to track the falling object had picked up a heat signature in the forest at exactly the spot where the airmen said they'd seen the glowing object. Accordingly, the base had gone on full alert. Later that morning after sunrise, security teams returned to the forest, where they found landing-gear impressions in the dirt, apparently left by the object they'd seen during their earlier encounter.

Events repeated themselves, but with a much greater intensity in the late-night darkness of December 27, 1980. The mystery lights returned, and deputy base commander Lt. Col. Halt and other officers led separate security teams into Rendlesham Forest to investigate. One of the first security patrols to reach the light came upon a truly eerie scene: they could see the light changing colors as it glowed through the trees. The closer they approached, the more they became enveloped in a thick yellow ground fog that clung to the forest floor and swirled around the men's knees. As they entered a clearing, the object again reacted to their presence by increasing its glow and floating away.

Then the object emitted a pulse of energy that shattered the silence of the night. Air Force personnel were enveloped by the energy burst, and each man could feel his hair stand on end from a tremendous charge of static electricity. Over their radios, the men were ordered to withdraw from the clearing and reinforce Col. Halt's detail, which was approaching from another direction with powerful "light-alls" that were to be set up around the object.

By the time Halt's detail had cordoned off the clearing, more security personnel had arrived and the area was now well lit. As a staff car arrived on the scene with senior officers, an orb-like object seemed to emerge from the back of the first object and floated a few feet above the ground. Security teams stood in awe as they observed what appeared to be creatures inside the floating orb.

Witnesses later testified that the orb floated in front of the base's senior officers, and the child-sized creatures inside, wearing silvery jumpsuits, appeared to be communicating with them, though no one could hear any words being exchanged. Security officers ordered the enlisted personnel manning the cordon to disperse and track other falling lights now dropping into the forest.

Whatever transpired between the officers and whatever was inside the orb is unknown. The only official written documentation of the Bentwaters event was recently released by the British Office of Official Secrets, and researchers believe the documents had been sanitized. The 1980 Bentwaters case still remains an enigma. Was it really an encounter between the military and extraterrestrials, or was it an elaborate security rehearsal using non-lethal psychological weapons to test the response of military teams at one of NATO's most important and secure air bases?

Berkelium

Number 97 on the periodic table of elements, with an atomic weight of 247, berkelium is reported to be the substance that powers flying saucers. In one crop circle design in a Kansas grain field in 1991, witnesses reportedly saw the equation *E 97+* in the sheaves of grain. Shortly after this equation appeared, it was eradicated by agents from the U.S. government, or by personnel who said they were government agents. This, of course, fueled theories that the government is still involved in removing as many traces of physical evidence of UFOs as it can.

Berkener, Lloyd V.

According to nuclear physicist and UFO researcher/author Stanton Friedman, Dr. Lloyd V. Berkener was a member of the super-secret UFO advisory group called MJ-12. Berkener was an explorer and scientist, a member of the CIA panel that determined that UFOs did not constitute a threat to U.S. national security, the executive director of the Carnegie Institute, and a member of the Joint Research and Development Board in 1946, a year prior to the Roswell crash in 1947. Berkener's presence on MJ-12 would not have been surprising, considering his credentials. *See* **Friedman, Stanton; Majestic**

Berlitz, Charles

Author of numerous books on UFOs, including *The Roswell Incident: The Classic Study of UFO Contact* with William L. Moore, among many more otherworldly topics.

Bermuda Triangle

Long thought to be a realm of intense UFO activity, the Bermuda Triangle—a section of the Atlantic Ocean bounded at its points by Melbourne, Florida; Bermuda; and Puerto Rico—is a region where ships and planes have mysteriously vanished without a trace. Along with UFO and USO (unidentified submersible object) sightings, there have been reports over the years of bizarre magnetic and atmospheric phenomena that render ship and aircraft compasses and engines useless, and some researchers believe the area to contain the remnants of the lost continent of Atlantis.

The Triangle was also the site of the famous 1945 disappearance of Flight 19, in which five U.S. Navy planes, piloted by inexperienced aviators, and a rescue plane sent to locate them, all disappeared without a trace. According to Navy records:

> At about 2:10 P.M. on the afternoon of 5 December 1945, Flight 19, consisting of five TBM Avenger Torpedo Bombers, departed from the U.S. Naval Air Station, Fort Lauderdale, Florida, on an authorized advanced overwater navigational training flight. In charge of the flight was a senior qualified flight instructor, piloting one of the planes. The other planes were piloted by qualified pilots with between 350 and 400 hours flight time of which at least 55 was in TBM type aircraft. The general weather conditions were considered average for training flights of this nature.
>
> A radio message intercepted at about 4 P.M. was the first indication that Flight 19 was lost. This message, believed to be between the leader on Flight 19 and another pilot in the same flight, indicated that the instructor was uncertain of his position and the direction of the Florida coast. The aircraft also were experiencing malfunction of their compasses. Attempts to establish communications on the training frequency were unsatisfactory due to interference from Cuban broadcasting stations, static, and atmospheric conditions. All radio contact was lost before the exact nature of the trouble or the location of the flight could be determined.
>
> Indications are that the flight became lost somewhere east of the Florida peninsula and was unable to determine a course to return to their base. The flight was never heard from again and no trace of the planes were ever found. It is assumed that they made forced landings at sea, in darkness somewhere east of the Florida peninsula, possibly after running out of gas. It is known that the fuel carried by the aircraft would have been completely exhausted by 8 P.M. The sea in that presumed area was rough and unfavorable for a water landing. It is also possible that some unexpected and unforeseen development of weather conditions may have intervened, although there is no evidence of freak storms in the area at the time.
>
> All available facilities in the immediate area were used in an effort to locate the missing aircraft and help them return to base. These

> efforts were not successful. No trace of the aircraft was ever found even though an extensive search operation was conducted until the evening of 10 December 1945, when weather conditions deteriorated to the point where further efforts became unduly hazardous.
>
> One search aircraft was lost during the operation: A PBM patrol plane, which was launched at approximately 7:30 P.M., 5 December 1945, to search for the missing TBMs. This aircraft was never seen nor heard from after take-off. Based upon a report from a merchant ship off Fort Lauderdale which sighted a "burst of flame," apparently an explosion, and passed through on oil slick at a time and place which matched the presumed location of the PBM, it is believed this aircraft exploded at sea and sank at approximately 28.59 N; 80.25 W. No trace of the plane or its crew was ever found.

Referred to by the Navy as *The Lost Patrol,* it was the subject of a book by Larry Kusche called *The Disappearance of Flight 19.* Although a tragic occurrence, the book suggests the disappearance of Flight 19 and its would-be rescuers was in no way paranormal and probably had more to do with bad weather and a series of inadvertent navigational errors than with UFOs.

An article published in *UFO Magazine* in 2002 reports on the findings of marine geologists who have studied the seabed beneath the "Triangle" and find it to be geothermally active, containing vast pockets of trapped methane-hydrate gas which can sometimes escape and reach the surface. The gas "envelope" can spread over hundreds or thousands of square yards of ocean surface and into the atmosphere and is believed powerful enough to cause passing ships to lose their buoyancy and sink, and is combustible enough to explode upon contact with a passing aircraft's engines.

This region of the Atlantic Ocean is also known for its "flash" squalls and "rogue" waves (which can sometimes reach a hundred feet in height), that can send an unlucky sailor to his doom. Whether or not the Bermuda Triangle is a natural or paranormal trap for the unwary, or merely another creation of the modern tabloid media, its legend lives on.

Bible, The

Many researchers in the UFO community believe that events depicted in the Bible, particularly in the Old Testament, are actually a retelling of historical fact: the implantation of the earth by creatures from another world to whom human beings are related genetically either by direct interbreeding or by active hybridization, which continues to this very day. Stories of the "giants in the earth," the Nephilim who married the descendents of Adam and Eve; the awesome energy of the Ark of the Covenant; the explosions that destroyed Sodom and Gomorrah; and Ezekiel, who saw a wheel "turning like a fire in the sky," are all believed by some researchers to be true accounts of UFO activities and encounters.

Big Whack Theory

One among many theories explaining the creation of the moon. The "Big Whack" presupposes that the moon was created when, at an early stage of earth's coalescence into a planet, an asteroid hit the earth and sent huge amounts of dust and debris into space where it was trapped by earth's gravity and itself coalesced into the moon. This also, of course, presupposes that at one time the earth and the moon were one celestial body, which, based upon what the Apollo astronauts brought back from the lunar surface, might not be true. *See* **Moon, Dark Side of the**

Billy the Kid

Although on the surface it would seem so remote as to be impossible that there could be any link between 1870s gunfighter Billy the Kid and the 1947 crash of a UFO outside of Roswell, New Mexico, there is indeed a very real connection.

Billy the Kid, *aka* William Bonney, *aka* Henry McCarty, fought in the Lincoln County wars and ran afoul of Sheriff Pat Garrett, from Roswell, who tracked him down and killed him on July 14, 1881. At least, that's what the history books say. However, there is a local legend, in part told by the remaining members of the Brazel family, that Pat Garrett never killed Billy the Kid. Instead, Garrett escorted him to the border, where William Bonney crossed into Mexico and spent the rest of his days as a retired gunfighter.

More important for ufology, however, is the fact that Pat Garrett, as the family legend continues, was investigating the Brazels for cattle rustling when he was ambushed and killed in 1908. One of the gunmen in that ambush was a young Mac Brazel, the very same Mac Brazel who discovered the Roswell debris in 1947. Therefore, some UFO researchers have naturally wondered just what it was that the

Army might have threatened Brazel with when they stuck him in jail for a week after he turned over the wreckage. Possibly, the Army held over Brazel's head a murder charge from 1908. Since there is no statute of limitations for murder, especially when the murder victim is a peace officer, this accusation could have sent Brazel to the gallows. *See* **Brazel, William "Mac"**

Birdsall, Graham W.

Graham W. Birdsall

Graham W. Birdsall was the editor of *UFO Magazine* in the United Kingdom and was one of England's most important UFO researchers. A figure on UFO documentary videos as well as the moderator at UFO conferences, Birdsall helped establish the legitimacy of research into the manifestations of crop circles, alien abductions, and such incidents as the reported UFO sightings and encounter at the military site RAF Bentwaters in 1980.

Birdsall, who died from a brain hemorrhage on September 19, 2003, brought serious UFO research into the public domain not only in England, but, through his magazine, appearances at conferences, and on videos, inspired UFO enthusiasts and researchers in Europe, the United States, and in the Third World.

"Black Arm, The"

A phrase attributed to former Kansas senator and Republican Majority Leader Bob Dole, to describe what he called a highly secret arm of the government's security branch, supposedly managing classified government projects. Dole referred to this group when, in a meeting with UFO research groups, he explained that Congress had been trying to get the real UFO information out to the public, but that they were being consistently blocked by what he termed "the black arm of government."

He might have been referring to one or more groups under the control of the National Security Agency, itself a group that came into being in 1947 after the passage of the National Security Act. The "black budget" and operations "going black" meant being buried under the security blanket of the National Security Agency and out of Congressional oversight.

Black Helicopters

These aircraft, if not an urban myth, are a mechanized version of the "Men in Black." In reality, specially modified Army helicopters with special noise reducers on their rotors and engines have been developed for urban warfare and intelligence operations and have been involved in various secret training missions.

Local police, who are usually unaware of these missions, have actually fired on Army troops rappelling from helicopters because they believed they were either terrorists or burglars. Abductees, contactees, and people who have had UFO encounters often report being followed by black helicopters. It's unknown whether the helicopters are ours or are piloted by alien-controlled humans who are keeping tabs on those who have had UFO encounters. *See* **Military Abductions (MILAB)**

Blair, William

A scientist at the Boeing Institute of Biotechnology who has discovered that "obelisks" on the moon's surface form six isosceles triangles, a geometric figure whose naturally occurring creation would be next to impossible. This points to the possibility that the obelisks might have been artificially created by intelligent beings who might have had a presence on the moon long before we got there. *See* **Moon, Dark Side of the**

Blanchard, William

Commanding officer of the 509th Bomber Squadron at the Roswell Army Air Field in 1947. Col. Blanchard ordered his Public Information Officer, Lt. Walter Haut, to release a statement to the newspapers that a team from the air field had retrieved a "flying disc" that had crashed on a ranch outside of Roswell, New Mexico. Blanchard had learned of the crash from his intelligence officer, Maj. Jesse Marcel, who had learned of it from Chaves County Sheriff George Wilcox, who had learned of it from rancher Mac Brazel, who found the debris. Under later orders from Eighth Air Force commandant Gen. Roger Ramey, Blanchard retracted the original press release and issued the weather-balloon cover story. And that is how the official government cover-up of UFOs began.

William Blanchard

Blue Book

See Projects, Government UFO

Blum, Howard

A *New York Times* writer and author of *Out There,* a book which shows that government officials—particularly the U.S. military and science advisers with security clearances above Top Secret—have formed various study groups over the years while denying their ongoing interest in the phenomenon.

Boas, Antonio Villas

This early case is significant to UFO researchers because it was reported long before such abduction stories were known to the public. Boas was a young Brazilian farmer at the center of this 1957 encounter. Twenty-three-year-old Boas was tilling the fields with his tractor early on the morning of October 16 when he noticed what seemed to him to be an extremely bright red star. The "star" began to approach the young man, and as it got closer, Boas could see it was actually a craft of some kind that suddenly projected three legs and landed in the nearby field. The lights from the craft were so bright they completely washed out the light coming from the tractor.

Scared to death, Boas stepped on the gas, but the tractor's engine died. He jumped off and tried to run, but was grabbed by a humanoid-like creature and held tight. As Boas tried to fight off the attack, three more creatures surrounded him, somehow paralyzed him, and dragged him by his feet into the craft. Once inside, Boas was taken to a small room, where his clothes were taken from him and the beings coated his bare body with a kind of gel. According to Boas, the beings wore tight-fitting spacesuits complete with helmets, had finely featured faces with small blue eyes, and made strange sounds that sounded like barks or yelps.

Boas was then led into another room, which had strange red letters written over the doorway, letters whose shapes he memorized and was later able to write down. In the room, a blood sample was drawn, and he remembered that there was an odd odor in the air that made him feel nauseous.

Shortly thereafter, a young woman entered the room. She was about five feet tall, according to Boas, and very beautiful. The woman had an angular face with eyes that were catlike, and long white hair. She was nude and Boas noticed that her pubic and arm hair were a strange, bright red. He became aroused and the two had intercourse, which, though more than pleasant, appeared to have been the main purpose of his abduction.

The woman also seemed to be relieved when the whole process was over, as if this was something she was more obligated than anxious to do. Before leaving, the woman pointed to her abdomen and to the ceiling, making other gestures meant to convey to Boas in no uncertain terms that he had fathered her just-conceived child. Angered, he said later that it was clear that "all they wanted was a good stallion to improve their stock." The beings then gave him a courtesy tour of the ship and dropped him off. Four hours had passed since he had been picked up.

Boas eventually became a lawyer with four earth children of his own. To this day he can still remember every detail of his encounter even without the use of hypnotic regression.

Boisselier, Brigitte

Brigitte Boisselier

The director of Clonaid, Dr. Boisselier is a French chemist who rocked the medical establishment in 2003 with the announcement that her organization, associated with the Raelians, had performed the first successful cloning of a human baby. Although Boisselier was unable to provide DNA evidence to support Clonaid's claim, she also announced there were other cloned children about to be born in Europe and Japan. Boisselier's pronouncements were of particular importance for the UFO community because of Raelian members' claims that cloning is the method extraterrestrials used to create the human species, hybridizing their own DNA with that of earthly DNA. *See* **Clonaid; Rael; Raelians**

Bolling Air Force Base

On July 19 and 20, 1952, radar controllers at Bolling Air Force Base were among the many military personnel who tracked UFOs over Washington DC while the Air Force was powerless to intercept them. *See* **Washington D.C., UFO Overflights**

Bonilla, A.Y.

Possibly the first person ever to take photographs of UFOs, Professor Bonilla was photographing the sun at the Zacatecas Observatory in Mexico in August 1883 when he captured a number of UFOs on a photographic plate. Bonilla said that he observed 283 bright luminous disks in the sky as well as 116 dark disks that moved across the face of the sun.

Borman, Frank

Gemini astronaut Frank Borman, along with astronaut James Lovell, reportedly saw a UFO during the second orbit of their fourteen-day flight in 1965. When Borman said he had spotted a UFO near their capsule, mission control said he was probably seeing the final stage of their own Titan booster rocket. Borman confirmed that he could see the booster rocket, but that he was watching something else, something completely different through the window of the space capsule. The radio transcript of their communication with mission control has been published but not commented upon by NASA:

> ***Gemini 7:*** *Bogey at 10 o'clock high.*
>
> ***Control:*** *This is Houston, say again 7*
>
> ***Gemini 7:*** *Said we have a bogey at 10 o'clock high.*
>
> ***Control:*** *Gemini 7, is that the booster or an actual sighting?*
>
> ***Gemini 7:*** *We have several, actual sighting.*
>
> ***Control:*** *Estimated size or distance?*
>
> ***Gemini 7:*** *We also have the booster in sight.*

See **Apollo Program; Lovell, James**

Bourgogne, Duke of

In his memoirs written in the fifteenth century, the Duke of Bourgogne wrote that on November 1, 1461, "an object appeared in the sky over France." The object, the duke observed, "was as long and as wide as a half moon; it hung stationary for about a quarter of an hour, clearly visible, then suddenly ... spiraled, twisted and turned like a spring and rose into the heavens."

Bower, Doug

One-half of the English duo who claimed to have created crop circles by tramping down the grain using wooden boards and string.

Although many people thanked Doug and Dave for demonstrating how they had hoaxed crop circles, the real story behind why these two farmers went public was never made known. Apparently, although Doug Bower and Dave Chorley were able to show how they could make impressions in grass, they never recreated a complete crop circle of the type that appear and continue to appear in England.

Also, it was pointed out that the real motive of Doug and Dave was not so much to show how they could hoax crop circles for fun; rather, they were trying to show that crop circles were hoaxes so that people would stop coming into the fields of private farmers to walk among the real crop circles. Apparently, tourists were causing so much damage to farmers' crops that Doug and Dave hoaxed the process just to keep the crowds away by making them think that crop circles were a complete hoax. However, there were crop circles long before Doug and Dave and crop circles continue long after Doug and Dave. *See* **Crop Circles; Chorley, Dave**

Boyd, Alpha

Daughter of Roswell civilian airplane mechanic Ervin Boyd. *See* **Children of Roswell**

Boylan, Dr. Richard

Researcher of hybridization of children by extraterrestrials, lecturer and author Richard J. Boylan is a behavioral scientist who uses hypnotherapy on patients attempting to recall details of their partially remembered close encounters and for exploring previous life experiences stored in subconscious memory. Dr. Boylan is vice-president of the Academy of Clinical Close Encounter Therapists (ACCET) an educational and research organization and has presented papers on his research at the 1992 MIT Abduction Study Conference and the 1995 Cosmic Cultures International Conference in Washington, D.C. He is author of *Close Extraterrestrial Encounter*, *Labored Journey to the Stars*, and *Project Epiphany*.

Braun, Wernher von

Receiving his Ph.D. in physics from the University of Berlin, von Braun was a German rocket scientist who developed the V-2 missile that was successfully launched against England towards the end of the war. Arrested and then released by the SS and the Gestapo and with a death sentence over his head, von Braun escaped from Peenemunde with a team of rocket scientists and surrendered to the Americans who, realizing what a find von Braun was, descended upon Peenemunde and Nordhausen and captured the V-2s.

Von Braun came to America after World War II via Project Paperclip, and along with other scientists taken out of Germany, worked at Huntsville, Alabama, where he oversaw the development of America's guided missile program.

Von Braun later moved to NASA and was the guiding light behind the construction of the Saturn 5 booster used to launch Apollo to the moon. Regarding UFOs, von Braun is quoted in a 1959 *News Europa* article as saying:

> We find ourselves faced by powers which are far stronger than we had hitherto assumed, and whose base is at present unknown to us. More I cannot say at present. ... We are now engaged in entering into closer contact with those powers, and in six or nine months time it may be possible to speak with some precision on the matter. This refers to mysterious events during the re-entry phase of a Juno 2 rocket during a test flight. He later went on to add that it is as impossible to confirm them [UFOs] in the present as it will be to deny them in the future.

Brazel, William, Jr.

Rancher Mac Brazel's son, who was shown pieces of the Roswell debris by his father and described it as: "something on the order of tin foil, except (it) wouldn't tear. You could wrinkle it and lay it back down, and it immediately resumed its original shape. It was pliable, but wouldn't break, weighed nothing, but you couldn't scratch it with your fingernail." Brazel said he was contacted by the military after his father was taken into custody and ordered not to talk about what he had seen. *See* **Children of Roswell**

William Brazel, Jr.

Brazel, William "Mac"

Manager of the Foster Ranch, who, according to most accounts, was the first to discover the debris field of the UFO crash at Roswell and bring it to the attention of the authorities.

On the morning of July 5, 1947, following a night of tremendous thunderstorm activity, Mac Brazel decided to check on his livestock. While riding his horse across the range, accompanied by seven-year-old William D. Proctor, he discovered a large amount of strange debris strewn across a pasture. The debris consisted of pieces of lightweight material that resembled tinfoil and light I-beams with unknown symbols engraved along their sides. The debris was so unusual that Brazel gathered up a bunch of it to take home.

The next day Brazel took a boxload of the debris to Chaves County Sheriff George Wilcox in Roswell, who put the wreckage inside one of his jail cells and called the air base. He spoke to Maj. Jesse Marcel, the Roswell Army Air Base intelligence officer. Marcel was ordered by Base Commander William Blanchard to go out to the Foster Ranch with Brazel to take a look around. It was nightfall when they arrived, so they spent that night in a small cottage and the next morning toured the debris field.

The Army was called in shortly thereafter and cordoned off the area. They retrieved the box of material Brazel had brought to Sheriff Wilcox, threatened the sheriff and his family into silence, and took Brazel into custody. They held him for a week, after which time he reappeared in Roswell with a new pickup truck and a new appreciation for silence. Brazel later told the local newspaper that he was sorry he had ever found the debris or mentioned it to the authorities. *See* **Billy the Kid; Roswell, New Mexico**

Brazilian Navy Case

While on training exercises in the South Atlantic Ocean near Trindade Island on January 6, 1958, forty-eight seamen, the ship's captain Carlos Alberto Bacellar, and several civilian observers on board the *Almirante Saldanha* all witnessed the flight of several flying saucers overhead.

Among the civilians was a technical photographer, Almiro Barauna, who took a series of photos of the craft cavorting in broad daylight. The film was developed immediately in a makeshift darkroom on the ship, with Captain Bacellar personally watching the entire development process to insure the film was not tampered with and to give added credibility to the sighting.

On February 21 the president of Brazil, Juscelino Kubitschek, stated that the Brazilian Navy had thoroughly analyzed the photographs and that he, as president, vouched for their authenticity, one of the extremely rare occasions where a head of state has endorsed the process of ascertaining the nature of unidentified flying objects. These "daylight disk" photos were to become the clearest proof yet of the reality of the saucer phenomenon, and they have never been debunked. *See* **Bacellar, Carlos Alberto; Barauna, Almiro**

Brett, Robin

A NASA scientist who, commenting upon the logical impossibilities surrounding the nature of the "hollow" moon with different centers of gravity below its surface, said, "It seems easier to explain the nonexistence of the moon than its existence." *See* **Moon, Dark Side of the**

"Bridge, The"

An apparently artificial twelve-mile-long structure on the moon's surface observed by *New York Herald Tribune* science editor John J. O'Neill, who reported his finding to the Association of Lunar and Planetary Observers. Although the association members laughed at O'Neill's "sighting," they stopped laughing when British astronomer Dr. H. P. Wilkens reported the same observation to the BBC. Wilkens said, "It looks artificial. It's almost incredible that such a thing could have formed in the first instance, or if it was formed, could have lasted during the ages in which the moon has been in existence." Another British astronomer, Patrick Moore, said that the Bridge had appeared almost overnight. *See* **Moon, Dark Side of the**

Bronk, Detlev

According to Stanton Friedman, Bronk was another reported member of the group known as MJ-12. Dr. Bronk was a physiologist, aviation expert, and prominent member of the National Academy of Science. He served as president of both Johns Hopkins and Rockefeller Universities, was chairman of the Nuclear Research Committee, and the medical advisor for the Atomic Energy Commission. Bronk also served on the Scientific Advisory Committee of Brookhaven National Laboratory, where he worked with Dr. Edward Condon, chairman of the Condon Committee on UFOs. *See* **Majestic**

Brookings Institution

Established on March 13, 1916 by a group of business leaders and academics as a nonpartisan private research agency in Washington, the Institute for Government Research—forerunner of today's Brookings—was intended to be an entity that would promote government efficiency through a critical analysis of government administration and operation. The IGR was chaired by Frank Goodnow, President of Johns Hopkins University. The vice chairman was Robert S. Brookings. Eventually Brookings became its chairman.

During the Depression, to stay financially viable, the Brookings Institution, as it came to be known, took contract research jobs from other organizations besides the government, even though the government continued to be one of its prime clients, relying upon its research for New Deal legislation and then mobilization at the outbreak of World War II. With postwar grants from the Rockefeller and Ford Foundations, the Brookings Institution expanded its scope, bringing in scholars from a variety of academic fields and reestablished a graduate fellowship program.

In the field of UFO research, the Brookings Institution played an important role by developing a report on the proposed Studies on the Implications of Peaceful Space Activities for Human Affairs for the House of Representatives Committee on Science and Astronautics. In this report, Brookings researchers assessed the implications of human discovery of extraterrestrial intelligent life. Suggesting that artifacts of just such an intelligent life form might one day be discovered on Mars, the moon, or even Venus, the report also hypothesizes that members of an intelligent and advanced species might visit earth one day. The impact of the discovery of just such a race would be "electrifying" to members of society, the report predicts, especially to members of fundamentalist sects which are "growing apace around the world."

Accordingly, the report suggests, one of the options the government might choose is to withhold the information about the discovery of extraterrestrial life forms from the population. Ufologists believe that this report may form the basis for part of the government

A promotional poster from a 1950s-era film which depicts the panic associated with UFO contact. This is the kind of panic that prompted the Brookings Institute to suggest that it was best if information about UFOs be kept from the public.

policy not to disclose information about UFOs and extraterrestrials to the American people. *See* **War of the Worlds**

Brooklyn Bridge Abduction Case

The famous Linda Cortile abduction case, reported by author Budd Hopkins in his book *Witnessed*, in which a Manhattan woman was spirited away by a UFO in broad daylight from one of the world's most populated cities.

On November 30, 1989 at about 3:00 A.M., a large flying saucer appeared over New York City. Two small beings floated Cortile (whose real name is Napolitano) out of her twelfth-floor apartment on a beam of light, passing her right through a closed glass window and into the saucer. There were several eyewitnesses to the event, including drivers of cars on the Manhattan and Brooklyn bridges, which were stopped cold when the power went out and their engines died.

Two of the witnesses were reportedly U.S. Secret Service agents who were escorting the Secretary General of the United Nations, Javier Pérez de Cuéllar. According to published reports, the Secretary General, along with other people on the bridges, were a totally captive audience, and many of them began to panic.

After picking up Cortile, the saucer dove into the East River and disappeared. Apparently, memories of the event were wiped completely from the minds of the witnesses, some of whom later underwent regression therapy to recover this "missing time."

In an interview with *UFO Magazine* in 2002, Yancy Spence, one of the witnesses to the abduction, recalled the events of that night and remembers that he saw the two aliens escorting Cortile through the window and into a saucer hovering overhead. When Cortile was returned to her home, she found that her husband and son had been "switched off," a state of suspended animation that is a common occurrence among family members left behind during an abduction. *See* **Abductions; Cortile, Linda; Hopkins, Budd; Napolitano, Linda**

Brown, Melvin E.

A soldier who, on the night of July 5, 1947, was a guard at the Roswell crash site. Brown was ordered to get into the back of a truck with the recovered alien bodies. Although he was ordered not to look at the bodies under the tarp, Brown lifted it up and saw that the dead creatures were small, with large heads, and had odd-colored skin.

Bulawayo Airport, Zimbabwe

In a well-recorded witnessed sighting, a flying saucer sped by the control tower on July 22, 1985. Dozens of people on the ground saw the UFO as well as control tower personnel. Two jet fighters were scrambled to intercept the craft, a saucer with a bubble on top, but the jets could not catch up to it.

Burt, Harold

Author and UFO researcher, Harold Burt is a regional director of MUFON. His book, *Flying Saucers 101,* is the source for some of the entries in this encyclopedia, and it is an excellent starting point for young people interested in learning more about the abduction and visitation issues of the subject. Burt also examines the ancient history of UFO sightings. *See* **Mutual UFO Network (MUFON)**

Harold Burt

Bush, Vannevar

An alleged member of MJ-12, Dr. Bush is also credited with developing the intellectual theory behind the entire World Wide Web and the hyperlinks that allow users to click their way from website to website. In a 1945 article in *The Atlantic Monthly,* Bush proposed that a hyper-linked methodology of pursuing information, the way one might jump from entry to entry in a dictionary, was probably a more intuitive and efficient way of searching for information than a strictly linear pursuit. Far ahead of its time, the article presaged the way users search for information on the Internet by using various types of hyperlink tagging systems such as HTML and XML.

A leader in research and development at MIT and Carnegie Institute, Dr. Bush was at one time the head of each of the following: the Office of Scientific Research and Development, the Joint Research and Development Board, and the National Advisory Committee on Aeronautics. Dr. Bush also put together the National Defense Research Council in 1941. *See* **Majestic**

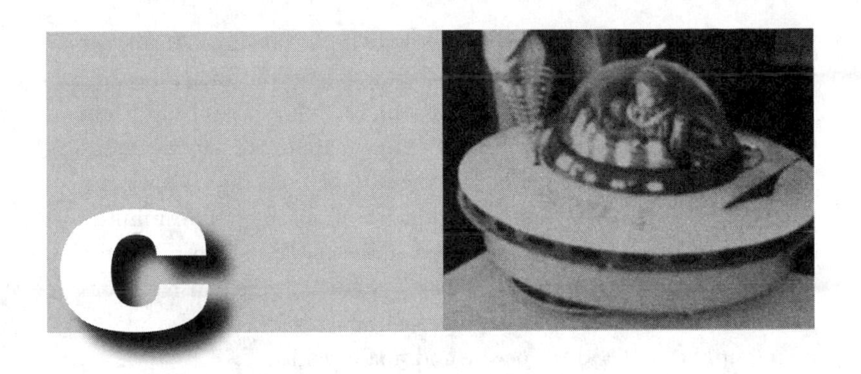

Cahill, Helen

In a signed affidavit, Helen Cahill, sister of Roswell witness Frankie Rowe, attested to the veracity of Rowe's story about handling material from the crashed Roswell craft and how her family had been threatened into silence by the military. *See* **Children of Roswell**

California, University of at San Diego (UCSD)

The metallurgy laboratory at the University of California San Diego tested material from a subcutaneous implant from an abductee that Dr. Roger Leir believed might have been of alien origin and found that the material composing the implant was of extraterrestrial origin. *See* **Implants, Alien; Leir, Roger**

"Calling Card" Plaque

A plaque placed on the Voyager spacecraft by NASA, which describes human beings and the location of planet Earth in the event that extraterrestrial beings encounter Voyager.

Canary Islands Case

A well-documented 1976 case, of which major local newspapers said: "Thousands of people have seen a spectacular luminous phenomenon." Eyewitnesses included civilians, a physician, priests, and engineers. The event was also seen by the entire crew of the Spanish naval ship *Atrevida*. The ship's captain filed this report:

At 21:27 hrs. on 22 June, 1976 we saw an intense yellowish-blue

> light moving out from the shore towards our position. At first we thought it was an aircraft with its landing lights on. Then it became stationary. The original light went out and a luminous beam from it began to rotate. It remained like this for approximately two minutes. Then an intense halo of yellowish and bluish light developed and remained in the same position for 40 minutes. Two minutes after the great halo vanished, the light split into two parts, the smaller part being beneath the part from which the bluish nucleus had come. The upper part began to climb in a spiral, its glow lighting up the land and the ocean, and finally vanished.

Later that evening in the nearby town of Las Rosas, a physician, Dr. Francisco Padrón, filed a formal deposition attesting to what he and other residents had seen. Dr. Padrón said that a large luminous sphere had hovered over the road, and:

> It was made of a totally transparent and crystalline-like material, since it was possible to see through it the stars in the sky; it had an electric-blue color but tenuous, without dazzling. It had a radius of about 30 meters, and in the lower third of the sphere you could see a platform of aluminum-like color as if made of metal, and three large consoles. At each side of the center, there were two huge figures of 2.5 to three meters tall (10 feet), dressed entirely in red and facing each other in such a way that I always saw their profile. Then I observed that some kind of bluish smoke was coming out from a semi-transparent central tube in the sphere, covering the periphery of the sphere's interior without leaking outside. Then the sphere began to grow until it became huge, like a 20-story house, but the platform and the crew remained the same size. It rose slowly and majestically, moving slowly toward Tenerife. Suddenly, it reached enormous speed like none I ever saw in an airplane. Then, it disappeared in the direction of Tenerife.

When Dr. Padrón witnessed this event, he was riding in a taxicab. The cab driver confirmed that he too saw the same thing. The driver stated: "A craft that looked as if it was made of transparent crystal, about 25 meters (85 feet) high and 20 meters (65 feet) wide, with two persons dressed in brilliant red inside."

In 1994 the official Spanish Air Force files on this case were declassified and released to the public in an ongoing effort to acclimate them to the existence of flying saucers. In the files were many more depositions confirming the events of June 22, 1976. There were also photographs of the blue transparent globe as well as hard radar readings, and a report from the naval photo lab attesting to the authenticity of the photographs. *See* **Padrón, Francisco**

Cape Canaveral, Florida

Launching facility for the U.S. space program. It was also called Cape Kennedy, but its name was changed back to the original Cape Canaveral and the launch facility itself was named the Kennedy Space Center. The Kennedy Space Center at Cape Canaveral is the manned mission NASA launch facility. See **National Aeronautics and Space Administration (NASA)**

Capitol Building

On July 10 and 20, 1952, UFOs buzzed the U.S. Capitol, the White House, and the Pentagon. U.S. Air Force jets were unable to intercept them or prevent them from returning. *See* **Washington, D.C., UFO Overflights**

Cardenas, Filiberto

An abduction case involving sightings of underwater UFOs. In the early evening of January 3, 1979, Cardenas and three of his neighbors were driving along a highway outside of Hialeah, Florida, when their car suddenly stalled, and the car's electrical system failed. After getting out of the car to investigate, Cardenas and his passengers heard a very strange loud buzzing sound. Suddenly a bright blue-violet beam of light fell over Cardenas and he was instantly paralyzed. The light came from an odd-shaped craft hovering above the car. The craft was dark with poorly defined edges, like a blurry picture.

The beam lifted Cardenas above the treetops in full view of Fernando Marti, his wife Elizabeth, and their thirteen-year-old daughter Mirta. All three people sat stunned, as the craft slowly flew off with Cardenas inside.

Cardenas turned up about two hours later, and told a very dramatic story. The craft, he said, had carried him for some distance when it suddenly dove into the ocean. Cardenas said he could see water rushing by through a nearby porthole. He felt and sensed that there was absolutely no resistance against the water, the craft cutting through it like a hot knife through butter.

The beings who had spirited Cardenas away were small, but looked very human. They communicated with him through telepathy in perfect Spanish. They wore tight-fitting, one-piece jumpsuits. The craft took Cardenas to an underwater base that the aliens said they maintained. Then Cardenas was returned.

Several weeks later he was telepathically directed by the aliens to go back to the location of his encounter. Cardenas asked if he could bring his wife, Iris, and they agreed. During this next encounter the aliens told Cardenas and his wife how important universal love was for the health of the planet and the advancement of mankind.

At the time of one of the Cardenas' encounters, hundreds of people at Miami International Airport reported observing a large ship and two smaller disk-shaped craft hovering nearby. When interviewed later, all the eyewitnesses confirmed the same time and location of the sighting. It matched the time and location of the Cardenas' encounter exactly.

Carlsberg, Kim

An abductee who has reported in her book *My Wildest Dreams* that she has been the victim of military abductions. *See* **Military Abductions (MILABS)**

Carpenter, John

A UFO researcher who has traveled around the world documenting cases of UFO sightings and encounters, the overwhelming majority of which are either ancient or historical.

Carter, Jimmy

The thirty-ninth president of the United States and recipient of the Nobel Peace Prize, Jimmy Carter claimed to have had a UFO sighting in Georgia at a religious conference while he was campaigning for the presidency against Gerald Ford in 1976.

He filed a NICAP (National Investigations Committee on Aerial Phenomena) report of his sighting, and after his election promised the American people that he would do everything he could to reveal the secret of UFOs. However, before he got the chance to fulfill his promise, according to some reports, an official from the National Security Agency "threatened" Carter with legal sanctions for revealing any classified UFO secrets. Carter did not say another word about UFOs during his term in office.

Cash-Landrum Case

Vicki Landrum and Betty Cash approximately two years after their encounter with a UFO in the skies over Dayton, Texas.

A UFO case notable for both the sighting and its aftereffects. It involved Betty Cash, then a fifty-one-year-old businesswoman; Vicki Landrum, then a fifty-seven-year-old restaurant employee; and Landrum's grandson Colby, then seven years old.

On December 29, 1980 they encountered a large, flaming diamond-shaped object hovering above the road in front of their car in Dayton, Texas. The interior of the car became unbearably hot, forcing them out, but soon Colby and Vicki Landrum returned to the vehicle out of fear. Betty Cash remained outside the automobile for ten additional minutes. The diamond-shaped object hovering overhead was accompanied by twenty-three helicopters that Cash and Landrum assumed to be military.

The trio of observers was initially affected mainly by the heat and the bright light, and they all developed headaches. During the night Colby vomited repeatedly and his skin turned red; the same malady affected Landrum. Cash fared even worse: large water blisters formed on her face and head and by morning her eyes had swollen shut.

The three witnesses continued to have severe nausea and diarrhea, and their health deteriorated rapidly. Cash was taken to a hospital where she was treated as a burn patient; it was the first of more than two dozen periods of hospital confinement for Cash.

MUFON investigator John Schuessler listed the following medical problems developed by the three witnesses: eyes swollen, painful and watery; permanent damage to the eyes; stomach pains, vomiting and diarrhea; sores and scarring of the skin, with loss of pigmentation; excessive hair loss over a several-week period, the new hair having a different texture than the old; loss of appetite, energy and weight; damage to fingernails and shedding of fingernails; increased susceptibility to disease; and cancer.

The Cash-Landrum case was unique in that there was detailed documentation of the injuries, specifically photographs, witness corroboration, and reports of the subsequent medical treatment. The case was one of those cited by the Sturrock Panel as deserving of further study. *See* **Sturrock Panel**

Cattle Mutilations

Head of a mutilated animal whose jaw displays an incision made with surgical precision.

In the last three decades there have been scores of reports from across the United States of domestic farm animals being butchered with surgical precision, their key glands excised and taken away. No perpetrator has ever been caught, and no satisfactory official explanation has ever been offered.

According to researcher Linda Moulton Howe, a "standard" mutilation exhibits one or more of the following characteristics:

- *Usually occurs with cattle, horses, or sheep.*
- *Jaw cut: A precision, surgical-like cut of surrounding flesh to completely expose the jaw bone and teeth. The tongue and surrounding glandular tissue has been removed.*
- *The sex organs, navel, and nipples have been surgically removed.*
- *The anus and surrounding glandular tissues have been surgically removed.*
- *No evidence of any blood, despite the severity of the procedures.*
- *No footprints of any kind leading to or from the animal.*
- *The animal is often found in the middle of a perfectly created circle.*
- *The precision of the cuts and lack of blood rule out predator kills.*

Complicating the issue has been the number of kills attributable to wild predators such as wolves and coyotes which have been lumped in with the mutilations. Many local officials have speculated that the mutilations are the result of cultists, devil worshipers, or pranksters. Probably a good number of mutilations can be accounted for in this way. However, there have been far more of these mutilations than one might probably realize. Thousands of cattle have been lost over the years since 1967 in unexplained or strange circumstances. Although many of these kills are explainable, about 50 percent are not, and this group fits the classic pattern of a cattle mutilation.

The existence of cattle mutilations was brought to public attention by Howe and her Emmy-winning 1980 film documentary, *A Strange Harvest*, the first work that connected cattle mutilations to UFOs. Howe concluded that animal mutilations appear to emanate from two sources: UFOs and U.S. government participation.

In her books, including *An Alien Harvest,* Howe tells of discussions she's had with military intelligence operatives who told her that the mutilations were in fact being done by aliens who were testing the animals for high levels of toxins that humans have put into the atmosphere. They also told her that the black helicopters were a ruse to cover up the alien involvement in mutilations because the aliens didn't want the local farmers to panic and would likely then believe it was the government engaging in secret tests. *See* **Howe, Linda Moulton**

CAUS

See **Citizens Against UFO Secrecy**

Cavero, Carlos Castro

Former Spanish Air Force commander of Spain's Third Aerial Region who told interviewer J. J. Benitez in 1976 that the nations of the world were currently working together in the investigation of the UFO phenomenon in the hopes that when they acquired more "precise and definite information," they would have information they could release to the world.

Cavitt, Sheridan

An officer with the Counter Intelligence Corps at the 509th Bomber Squadron at the Roswell Army Air Field in 1947. Capt. Cavitt accompanied Maj. Jesse Marcel to the Foster Ranch UFO crash site and allegedly was part of the team that retrieved the alien bodies from the site and brought them back to the Roswell Army Air Field.

Cavitt, a career military man, has remained mostly silent about his reported involvement in the events taking place at Roswell in July of 1947, usually denying the "extraterrestrial" angle.

Cayce, Edgar

Edgar Cayce

Known as one of America's most gifted and powerful psychics, Cayce was called "The Sleeping Prophet" because most of his predictions were accomplished while he was in a self-induced trance state. He made predictions about things in the future, such as the ultimate fate of California and a cataclysmic event that would wreak destruction upon New York City. It is predicted by many scientists that one day the western edge of California will slide into the Pacific after a major 8-plus earthquake breaks off part of the state.

New York, some of Cayce's followers have argued, did, in fact, suffer from a cataclysmic destructive force on September 11, 2001. Cayce also predicted that in 1958 scientists would discover the ray that was used on the fabled city of Atlantis. In 1958, scientists at the combined Columbia University/US Army research laboratory at Columbia were working on lasers, which, some have said, were weapons deployed by extraterrestrials, one of which was discovered in the crash at Roswell.

What few people don't realize is that Edgar Cayce helped President Woodrow Wilson during his final months in office after he had become almost completely debilitated by a stroke. In fact, some Cayce scholars have said that Cayce ran the White House with Mrs. Wilson, transmitting instructions from the badly stricken president, to the president's staff. For those to whom such a possibility may seem incredible, it was Joan Quigley who was President Reagan's "seer," working with Nancy Reagan during the president's second term.

Edgar Cayce's nexus with ufology, it has been said, is based upon his ability to remote view, to project himself psychically onto another plane so as to receive information not necessarily accessible through the traditional five senses.

CDC

See **Centers for Disease Control**

Center for the Study of Extraterrestrial Intelligence (CSETI)

Founded by Steven Greer, CSETI has uncovered over 150 eyewitnesses to UFOs and UFO encounters. Dr. Greer's list of witnesses is comprised of former military personnel, defense contractors, NASA employees, government officials, and former members of the international intelligence community. Many of these witnesses confirm an intentional governmental cover-up of UFO activity. Greer has said that members of Congress and even senior officials in the CIA have asked him to brief them on what UFOs are and where they're from. The cover-up is so compartmentalized and self-sustaining that different bureaus within the same branch of government don't share this information. *See* **Disclosure Project; Greer, Steven**

Center for UFO Studies (CUFOS)

An independent UFO research agency comprised of scientists, and created in part by Dr. J. Allen Hynek after the Air Force shut down Project Blue Book in 1969. CUFOS continues the work that Blue Book was supposed to have pursued, had it been a true investigatory organization. *See* **Hynek, J. Allen**

Centers for Disease Control (CDC)

The Centers for Disease Control received the implanted metallic object that was removed surgically from Jack Weiner and sent it to a military pathologist in Washington, D.C., where an Air Force colonel took control of the object, and it was never seen again. *See* **Allagash Affair**

Central Intelligence

The precursor organization to the Central Intelligence Agency. Admiral Sidney Souers was the first Director of Central Intelligence and a member of MJ-12, according to Stanton Friedman. *See* **Central Intelligence Agency; Majestic**

Central Intelligence Agency (CIA)

With the exception of the military, perhaps no single department of government has been more involved with UFOs, both as a phenomenon to be investigated and as a subject to be used for its own purposes, than the Central Intelligence Agency.

The CIA was established as the successor to Central Intelligence, whose first director was the alleged MJ-12 member Admiral Sidney Souers; its first official director under its new name was another MJ-12 member, Admiral Roscoe Hillenkoetter. Thus, according to UFO researchers, the CIA began under the leadership of two senior military officers who were at the beating heart of the U.S. government's investigation into the nature of UFOs and the potential threat those UFOs (and those who piloted them) might pose to the security of the United States.

Over the years the CIA has declassified and released a number of memos regarding some of its UFO case studies, and document searchers have discovered references to UFOs in other CIA documents. What these document disclosures and successful searches have shown is that the CIA has been intimately involved with the subject of UFOs since its very inception. Indeed, some would say that the government agenda regarding UFOs and their supposed threat to the underpinnings of our national security is what has driven our country's national central intelligence strategy since the earliest days of 1947.

Among the more important documents divulged pursuant to FOIA requests served on the CIA are:

1952 National Security Council memo. *Subject:* Flying saucers. Directing the CIA to research and solve the problem of unidentified flying objects.

1952 CIA memo. *Topic:* Flying saucers. Discusses the phenomenal increase in saucer reports. Includes the official CIA version of the first flying saucer sighting by Kenneth Arnold in 1947. *Conclusion:* Flying saucers do not represent a military threat.

1952 Memo to Director of CIA from CIA Assistant Director of Scientific Intelligence. *Subject:* Unidentified flying objects. The memo states that sightings of unexplained objects at great altitudes and traveling at high speeds in the vicinity of major U.S. defense installations are of such a nature that they are not attributable to natural phenomena or known types of aerial vehicles. The memo establishes UFOs as a "priority project throughout the intelligence and defense research-development-community."

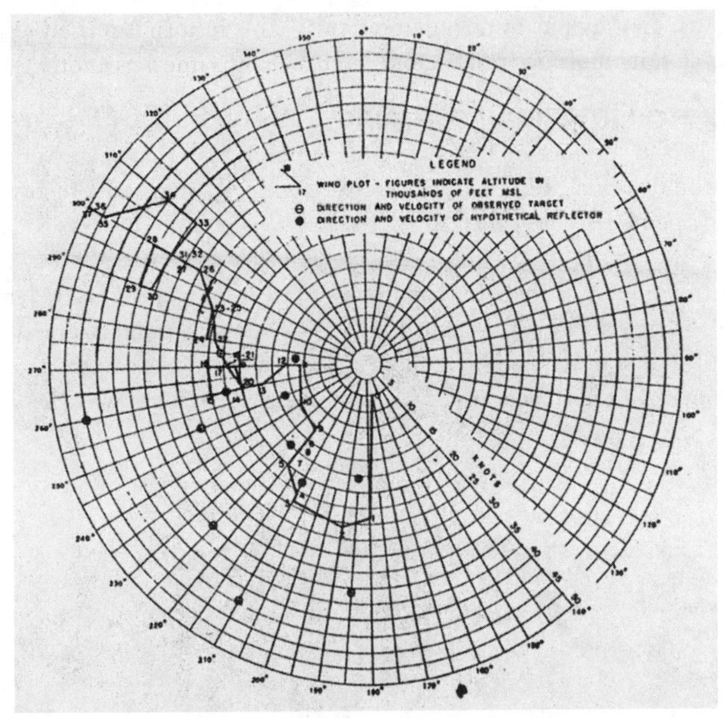

Radar Grid of UFOs over Washington, DC, 1952.

1956 CIA memo regarding a sighting in Eastern Europe. Reports that flying saucers were sighted flying in formation from Budapest to Moscow at an estimated speed of 12,000 kilometers per hour, at a time when top military planes could only fly at 600 miles per hour.

Shortly before Marilyn Monroe's death, the CIA reported an intercepted telephone conversation between Monroe and one of President Kennedy's aides, in which the actress threatened to reveal the secret of the UFO crash in New Mexico and the little "bodies" that were found in the wreckage. The CIA authorized the filing of the transcript of the intercepted phone conversation to Operation Bluefly and Operation Moondust, which, according to UFO researcher Clifford Stone, were actual operations authorized to catalog all information about downed extraterrestrial objects and the recovery of downed extraterrestrial objects.

In an intriguing communication from 1976, a CIA memo discusses the effects of magnetic and electromagnetic fields on the astronauts. The key line in this memo states: "This in turn is related to the possible propulsions of UFOs."

Central Intelligence Agency, History of UFOs in

In the summer of 1997 Gerald Haines, a historian then working at the National Reconnaissance Office (NRO), prepared a report for the open CIA publication *Studies in Intelligence.* Titled "CIA's Role in the Study of UFOs, 1947–1990," the report was apparently prepared at the request of then-Director of Central Intelligence (DCI) James Woolsey, after he was asked about UFOs during a radio interview.

The first of its kind, the report addressed the CIA's covert involvement with the subject of UFOs, admitting the reasons why the CIA

"The Flying Pancake." Air Force XF5U-2 disc plane c. 1947. A candidate for a flying saucer?

claimed no official interest in, or conducted serious investigation of, UFOs despite references to them in numerous memos. According to Haines' report, the underlying reason for the CIA's deliberate falsehoods concerning UFOs stemmed from Cold War fears that the Soviet Union could use UFO sightings as a means of overloading U.S. early-warning defense systems so that they could not distinguish them from real Soviet bombers. According to Haines, officials also feared that Soviet interference might touch off mass hysteria and panic among American citizens. A third reason dealt with more realistic national security concerns; America's then brand-new spy-planes were beginning to overfly Russian air space.

According to Haines, after the first "saucer" reports broke in the news, the Technical Intelligence Division of the Air Materiel Command based at Wright Field assumed command of Project Sign in 1948. Haines claimed that the Air Force determined that UFOs were real but easily explainable and not extraordinary. But the Air Force also recommended that the military continue to control investigations and did not rule out an extraterrestrial explanation.

In 1952 sightings seemed to reach a critical mass when UFOs "invaded" air space over Washington, D.C. This event alarmed the Truman administration and galvanized the country. The CIA reacted to the events by forming a special study group within the Offices of Scientific Intelligence and Current Intelligence to review the situation. Edward Tauss, acting chief of OSI's Weapons and Equipment Division, claimed that although most UFOs could be easily identified, investigations should continue, and the CIA's interest should be concealed from the public "in view of their probable alarmist tendencies."

DCI Walter Bedell Smith wanted to know what national security implications could be attached to UFOs and what use could be made of the UFO phenomenon in connection with U.S. psychological warfare efforts. Consequent to Smith's order for a review of national security information, the CIA met with the Air Force at Wright-Patterson and reviewed their data and findings. Air Force data suggested that 90 percent of all sightings were explainable, while the remainder involved incredible reports from credible people. The Air Force rejected theories that these sightings might involve secret weapons or aircraft from either the U.S. or U.S.S.R., or that they were "Men from Mars"—there was no evidence to support such ideas. Still lingering, according to Haines, was the fear of system overload and Soviet exploitation of UFO reports, preventing American forces from being able to distinguish real targets from phantom UFOs.

This particular danger was considered so significant by H. Marshall Chadwell, Assistant Director of OSI, that he recommended it be brought to the attention of the National Security Council in order that an agency-wide effort be made toward finding a solution. Chadwell briefed the DCI on the subject in December of 1952. He drafted a memorandum from the DCI to the NSC, and proposed an

NSC Directive be issued, establishing the investigation of UFOs as a priority project throughout the intelligence and defense research and development communities.

The Intelligence Advisory Committee was presented with the DCI's request that it informally discuss the UFO subject. The Committee agreed that the DCI enlist the services of selected scientists to review and appraise the available evidence and draft an intelligence directive on the subject. Maj. Gen. John A. Samford, Director of Air Force Intelligence, offered the Air Force's full cooperation. Thus was born the infamous Robertson Panel, charged with reviewing the available evidence on UFOs and weighing the possible dangers to national security.

Meeting from January 14–17, 1953, the Robertson Panel reviewed Air Force data on UFO case histories. After spending half a day studying the phenomena embodied in the data, they unanimously declared that explanations could be found for most if not all of the reported sightings, and that there was no threat to national security from UFOs. The panel also came to the conclusion that continued reporting of UFOs could threaten "the orderly functioning" of the government by clogging the channels of communications, or by "hysterical mass behavior" harmful to "constituted authority."

To meet these perceived problems, the Panel recommended that the National Security Council debunk UFO reports and institute a policy of public education to reassure the public of the lack of evidence behind UFOs. It suggested using mass media, advertising, schools, and even the Disney corporation to get the message out. The Panel also recommended that UFO groups such as the Aerial Phenomena Research Organization (APRO) and the Civilian Flying Saucer Investigators be monitored for "subversive activities." It was also noted by Haines that CIA officials demanded that agency interest in flying saucers be very carefully restricted; that the Robertson Panel findings be classified and no mention of the CIA sponsorship of the panel ever be released.

Haines reports that as of November 1954, the CIA and Lockheed's Advanced Development facility (the "Skunk Works") in Burbank, California, entered into a partnership to develop an overhead reconnaissance system for spying on the Soviet Union. By August 1955, the Agency was test-flying the U-2 aircraft, which could fly at altitudes above 60,000 feet, far beyond what military and commercial aircraft were then capable of. Yet Haines claims that this was a time-frame when most commercial pilots and air-traffic controllers began reporting more UFOs, the implication being that their sightings were misidentifications of U-2 flights.

The Air Force's own Project Blue Book records dispute this conclusion. The Haines report nevertheless suggests that 50 percent or more of all UFO sightings from 1955 through the 1960s were actually secret tests and flights of the U-2 and SR-71 Blackbird aircraft. The report also claims that the CIA conspired with Project Blue Book personnel to identify these aircraft sightings as types of atmospheric phenomena such as temperature inversions and ice crystals. Again, Blue Book records conflict with the Haines report. From records kept by the Fund for UFO Research, only about 2 percent of sightings were attributed to such atmospheric phenomena.

According to CIA critic Ralph McGehee, the agency's public affairs office churns out propaganda on a regular basis. "There was a document released by the Agency under the provisions of FOIA that talked about the activities of the Public Affairs Office," McGehee told *UFO Magazine* in an interview. "It said that [the] CIA had several thousand contacts with members of the media, and via their efforts [with this media], they had been able to repress criticism, turn intelligence failures into intelligence success stories, and other points and information they wanted to accomplish."

One of the first cases to draw the CIA fully into the controversy over UFOs was the Washington, D.C. overflights. According to Haines' report, "A massive buildup of sightings over the U.S. in 1952, especially in July, alarmed the Truman administration. On 19 and 20 July, radar scopes at Washington National Airport and Andrews Air Force Base tracked mysterious blips. On 27 July, the blips reappeared. The Air Force scrambled interceptor aircraft to investigate, but they found nothing."

The Haines report and UFO debunkers ignore the fact that these UFOs were seen on radar scopes all around the D.C. area, and witnesses on the ground and commercial airline pilots in the air saw and reported these objects. The report also fails to mention that during the second weekend "saucer invasion," an interceptor aircraft tried to chase these objects, but even flying at over 600 miles per hour, it failed to gain on the objects. Simply stated, Haines' conclusion that "no interceptors observed anything" is flat-out wrong, and many of his reported sources, such as Philip Klass, are at least highly suspect.

Regarding the disinformation strategy that Haines describes to characterize U-2 sightings as UFOs, it's a well-known fact that by 1960–61, the whole world was aware of the high-level spy-plane project. CIA pilot Francis Gary Powers had been shot down over the Soviet Union. The Russians were aware, and had been for some time, of the overflights of their country.

Perhaps more importantly, neither the U-2, nor the later SR-71, could make instant right-angle turns, hover, or come to treetop level, all characteristic behavior of UFOs reported again and again by witnesses. (Spy planes were tasked to fly in a straight line for hours, photographing installations at an altitude and speed that would make it almost impossible for U.S. adversaries to shoot them down.) Until the downing of Powers (and what Lt. Col. Philip Corso once described as a deliberate effort on the part of the CIA and the Air Force to trigger the Soviet Union's missile defenses in order to find out how they tracked a target), U-2s flying at 60,000 feet were virtually invisible; thus, there was no need to camouflage them as UFOs.

Haines concludes his report with this final irony:

> Like the JFK assassination conspiracy theories, the UFO issue probably will not go away soon, no matter what the Agency does or says. The belief that we are not alone in the universe is too emotionally appealing and the distrust of our government is too pervasive to make the issue amenable to traditional scientific studies of rational explanation and evidence.

The Haines report seemingly takes pains to document cases that contributed to public distrust of the CIA. This is a thick and thoroughly documented report, yet it's also a straightforward public admission by the CIA and Air Force that they perpetrated their own respective disinformation strategies upon the American public about UFOs. *See* **Corso, Philip; Disinformation; Klass, Philip; Project, Government UFO; Robertson Panel; Skunk Works**

Cernan, Eugene

Astronaut, commander of Apollo 17. Cernan was quoted in a *Los Angeles Times* article from January 6, 1973, as saying, "I'm one of those guys who's never seen a UFO. But I've said publicly I thought they were somebody else, some other civilization."

Chadwell, H. Marshall

Assistant director of Scientific Intelligence, who sent a memo to the Director of the CIA in 1952 stating that the sightings of UFOs over U.S. defense facilities were of such a nature that they could not be either "natural phenomena or known types of aerial vehicles." *See* **Central Intelligence Agency, History of UFOs in**

Challenger

The space shuttle Challenger exploded in 1986, killing all on board. Although Dr. Richard Feynman dramatically demonstrated that the cause of the explosion was due to a frozen O ring, there are still those who believe that Challenger was deliberately exploded by the Soviets who had employed a scalar energy-beam weapon against it.

Chassin, Lionel M.

The commanding general of the French Air Forces and General Air Defense Coordinator of Allied Air Forces for Central Europe, NATO who wrote that "mysterious objects have indeed appeared and continue to appear in the sky that surrounds us" in the foreword to the 1958 book *Flying Saucers and the Straight Line Mystery.*

Chatelain, Maurice

Former chief of NASA communications systems. In 1979, Chatelain wrote about the story of Apollo 11 astronaut Neil Armstrong having seen two alien spacecraft sitting on the rim of a crater when he was walking on the moon. Chatelain said, "The encounter was common knowledge in NASA, but nobody talked about it until now." *See* **Moon, Dark Side of the**

Chavez, Sam

Sergeant Sam Chavez of the Socorro, New Mexico, police department who received a frantic all from patrol officer Lonnie Zamora on April 24, 1964, begging him to get out to where he parked near a UFO with alien beings standing beside it. By the time Chavez got to Zamora's location, the craft had taken off, but the Air Force later confirmed that the landing was real because of the impressions made in the dirt. *See* **Zamora, Lonnie**

Cherwell, Lord

Friedrich A. Lindemann, later First Lord Cherwell, personal assistant to Winston Churchill and British Secretary of State for Air, who received a memo from Churchill in July 1952, asking if the reports of flying saucers could possibly be true. Lord Cherwell responded that they could probably be explained away as natural occurrences. *See* **Churchill, Sir Winston**

Children of Roswell

The children of the Roswell witnesses and some of the witnesses themselves to the UFO crash landing spoke to *UFO Magazine* in 1998 about the events they saw back in 1947. The combined weight of their testimony adds credibility to claims that it was a UFO that crashed that July, and the military engaged in great efforts to cover it up. These are some of their stories:

Alpha Boyd

In 1971, Alpha Boyd's father lay on a hospital bed recovering from a serious illness. As she watched him with concern, he began to tell her about the UFO that crashed in Roswell in 1947, even though he, too, had been threatened into silence.

> I've been too scared to talk about my father's story, but here it is. His name was Ervin Boyd, and he worked as a civilian airplane mechanic at the Roswell army base in 1947, out at a place called Hangar 84. In 1971 he had a light stroke. I came to Roswell to help care for him after his discharge from the hospital. One night he told me, "Honey, there's something that's been on my mind for a long time. I saw something a long time ago and was told not to say anything about it."
>
> "Daddy, what are you talking about?" I said, more concerned about him than what he was saying.
>
> "A spacecraft crashed here in 1947. I found out about it as I was taking a smoke break at the hangar. I had just stepped outside the hangar doors to light up when I saw a bunch of men and vehicles rushing to the hangar. I was a bit startled and wondered what was going on, when I noticed some men carrying what I first thought was the body of a child, and I asked myself why they didn't take it to the hospital. They passed right by me and I looked down and knew it wasn't something from this world. It was child-size, four feet, maybe a little more, and had a head that was larger than normal. The eyes were walnut-shaped and also larger than normal.

From my angle, it didn't look like it had much of a nose. The arms were a bit longer too, and the skin was ashy, gray, and kind of scaly. I don't know for sure, but I believe it was still alive when I saw it."

I would have been in shock at what I heard if I hadn't been worried about my father. He continued, "I asked one of the officers what it was. Suddenly, a bunch of other officers grabbed me and rushed me away from the area. One of them looked at me sternly, and said, 'Mr. Boyd, you did not see anything.'

"I demanded to know what was going on. They started getting rough with me, and then one of the officers I knew pretty well came to my rescue. 'I know this man, let me talk to him,' he said."

Dad said they threatened him not only with loss of his job, but also his life and the lives of his family. The next day, Dad went to the hospital to have them treat a minor injury he had suffered at the hangar. He asked the nurse that treated him if she knew what was going on.

Main Street in Roswell, New Mexico, c. 1947

"You shouldn't even be talking about that," she said, and abruptly changed the subject. The following day he returned for another treatment and asked for this nurse, but nobody knew where she was. He didn't know what happened to her, but thought they might have taken her away from the base.

I started asking questions and Dad appeared more nervous. "I probably shouldn't have said anything," he said, and stopped talking about it. After that, he changed the subject every time I tried to bring it up. Dad was a truthful man; well known and respected. He said it and I believe he told the truth. He passed away in December 1974.

I never asked my mother about any of this. Maybe I was too scared to talk about it, but I believe I might have come close to overhearing something. I remember hearing her and Dad whispering late at night. If I happened to walk in all the sudden, they would get real quiet. Mom never mentioned anything about this, even clear up to the day she passed away.

I've been scared to talk about it, and I'm not comfortable with it yet. Before I am, I'll have to sort out a lot of things and learn some more about the incident.

William Brazel, Jr.

Son of Mac Brazel, the ranch manager who discovered the wreckage on the Foster Ranch the morning after the crash.

William Brazel, Jr. in his naval uniform.

I was living in Albuquerque and picked up the *Albuquerque Journal* one morning. There, staring me in the face, was the image of my father, Mac Brazel. The article wasn't very complete, so I started asking around to find out what was going on. I was told he was being held at the Roswell base in protective custody.

The ranch was a big one—eighty sections, with seven hundred head of cattle and 3,500 to 4,000 sheep. A place like that won't hold together very long without someone looking after it. I left my job, jumped in the car, and drove to the ranch to help until my dad was released from the base.

If I recall correctly, he was released after three days. He was upset with the Air Force because he felt like he had been mistreated. How he got into that mess is that he saw Bob Scroggins one day in Carrizozo. Bob was an old friend of his and a state policeman. He told Bob what he had found on the ranch, and when he asked if he thought the Air Force should be contacted, Bob thought it best.

Dad had to go to Roswell on business, and he got in touch with the military while he was there. He did not take any of the material with him. The Army decided that he needed to stay there a while.

With Dad back at the ranch, I went back to my job in Albuquerque. Later that year, in November, I moved back to the ranch. Dad showed me where the object crashed. It was a hill with a very gradual slope, not high or steep. There was a mark where it hit and was dragged for about two hundred yards. You could tell it had to be traveling pretty fast when it hit the ground.

From the back of a horse, you can see a lot more than walking on foot. The few little bits and pieces I found were probably buried in the ground. A hard rain or wind storm sometimes would uncover it. They were more than splinters; maybe chips would be a better way to describe them. They were a light tan color. I found a piece of stuff I called tinfoil. It was more like lead foil, but it wasn't either one of them. There was an odd thing about the piece. When I found it, I wadded it up and put it in my chaps pocket.

Back at the ranch, I took it out of my pocket and laid it on the workbench for a few minutes, while I unsaddled my horse. When I looked at it again, it had flattened out. I tried it again, folding it this time, and watched it unfold. It took some time to do it, though. It was more the color of lead foil, which I was familiar with back in those days. There was also some stuff that looked like monofilament fishing line. It wasn't string and it wasn't wire. What's more, it wouldn't burn. The chips I found were like balsa wood, but you couldn't whittle or burn them. I kept all the stuff in a cigar box, so you know I didn't find very much, and it took six months to get that.

Mac Brazel discovered the crash debris at Foster Ranch.

People would ask if I ever found any pieces. I'd tell them maybe a few little scraps and that's all. People got to talking, I suppose. First thing I know, some Air Force officers showed up at the ranch, and they demanded the pieces I had found. They said it was a national security thing. I gave them everything I had, just like my Dad did.

They were real nice about it. No threats or anything. I didn't receive any threats until after I bought the place at Capitan and retired, so it must have been after 1985. I got a phone call and the guy said he was FBI. He warned me not to talk about this thing to anybody. He said they could put me in prison if I did. Made me pretty mad. He said he and his partner were coming to see me. "Never mind, don't come," I said. They never did. They might have come up to Capitan and checked me out.

I never thought the crash would become so famous and that so many people would seek me out. I'm sick and tired of it. The phone rings at any hour of the day. In fact, if the person on the line says they're calling about the UFO, I immediately hang up. It's been a nuisance.

Glenn Dennis

Glenn Dennis was a Roswell mortician who claimed that after the retrieval of the UFO debris from the crash site, the Army asked him for a number of baby-sized coffins. Dennis has continued to participate in UFO research, even though there are those in the UFO community who believe his story is fabricated.

I was put in charge of the funeral home's military contract work, the ambulance and mortuary service for Roswell Army Air Field. I first became involved in the event, now known as the "Roswell Incident," when I received a phone call from the air field Mortuary Officer on July 7, 1947. I don't remember his name, as they never had the same officer each day. He asked if we had any baby caskets, three and a half or four feet, hermetically sealed.

I told him we always kept a couple in stock. He wanted to know how long it would take to get more, so I told him by six the next morning. I asked him what was going on, and he gave me some excuse about "in case something happens in the future, we might need a lot of them." I didn't think too much about it until later, as we had that kind of inquiry often.

About forty-five minutes later, the same man called back. This time he wanted to know about embalming fluid; what chemicals it contained, what it would do to bodies that had been lying out in the open, would it change the stomach contents, would it change the tissue, the blood? He also wanted to know the procedures for removing bodies from a site and the preparation of bodies that had been lying out in the elements, that might have been shredded by predators. He also wanted to know if they transported a body under those conditions, without embalming, and how they should do it.

Back in those days we didn't have an air-conditioned hearse or a pathologist in Roswell. I told him I would go to a nearby dairy and buy all the dry ice I could and pack them in it. I also told him if he had a "hot one," that is, if he didn't know the cause of death, they had better contact a pathologist and do what he told them. I think I suggested they try Walter Reed Hospital in Washington, D.C., because I remembered bodies of local boys who died in the service coming through there. I also remember telling him, very politely, "You give us the specifications, you tell us how you want the bodies prepared, and we will prepare them according to your specifications."

An hour later we received a call for an ambulance for an airman who had been hurt on a motorcycle. I drove him to the base and went directly to the infirmary. When I swung into the driveway, there were three old Army field ambulances backed up at the ramp where I usually parked, and two MPs standing in between, so I drove around to the end and parked. The airman and I got out, and we walked up the ramp behind the ambulances.

When we got to the first ambulance, one of the rear doors was open and when you're in the business, naturally you're going to look. I saw something that looked like half of a canoe leaning up against the side near the open door. It was standing on end, and I was very close to it. It was about three, maybe four feet high. All around the bottom of this thing, all over the floor, was a lot of wreckage. It was all sharp and, as best I can remember, it was like broken glass. Some of the pieces and the "canoe" looked like stainless steel

Mortuary where Glenn Dennis worked.

that had been put through high heat. It shaded from very shiny to pink, to red, to brown, then black. There were markings on the canoe-shaped piece, around the outer edge along the curve and down one side. They were about four inches high, darker than the background, and clearly were deliberately put there. The markings reminded me of Egyptian hieroglyphics.

I saw the same kind of wreckage in the second ambulance. The doors were closed on the third ambulance, so I couldn't see what was in it. The MPs didn't even look at me. The airman saw the wreckage, but he was more concerned about his injuries. I followed him into the infirmary.

Glenn Dennis

I started down the hall to the lounge area to get a Coke. There was a lot of commotion up and down the hall, a lot of officers milling around, but I didn't know any of them. There was this captain (I remember seeing his bars) leaning next to an open side door. I went up to him and said, "It looks like we had a plane crash. Do I need to go in and get ready for it?"

I had never seen him before. He looked at me and said, "Who the hell are you?" I remember this very well. He was real snotty. I told him I was from the funeral home. He said, "Don't move from here, don't take one step," then walked away. After a few minutes he came back with two MPs. He told them, "Get this man off the base. He's off limits. You drive him back to town, make sure he gets there." They started to walk me down the hall. They weren't roughing me up or anything. They were real nice.

We had only gone a few feet when a voice said, "Bring that S.O.B. back here." We turned around and there was this big redheaded captain, about six-three, with a real short crew cut and the meanest eyes I had ever seen, like the devil himself, looking at me. He had a black sergeant with him, holding a clipboard. Someone must have gone and gotten them. The captain came up to me and poked a finger in my chest and said, "Look, mister, you don't go into Roswell and start a bunch of rumors that there's been a crash. Nothing has happened here, you understand?" He kept poking me, and of course I was getting a little upset. I said, "I'm a civilian and you can't do anything to me. You can go to hell!" That's when he jabbed me again and said, "Somebody will be picking your bones out of the sand." Then the sergeant said, "Sir, he would make better dog food." So I popped him off too. Then the captain said, "Get the son of a bitch out of here."

The MPs started taking me back down the hall again. That's when I saw a nurse that I knew was assigned to the base infirmary. She looked up and saw me, and she screamed, "Glenn! Get out of here as fast as you can!" She was sobbing and gasping for air, as she went on across the hall, through another door. Two men followed her; they were gulping for air too, and looked as though they were about to throw up. I didn't see or smell anything strange, though.

We went straight back to the funeral home, and they warned me to stay away from the base for the rest of the day. I tried to call back to the infirmary, as well as the nurse's quarters, to find out what was going on, but I couldn't get an answer. I did not say anything to anyone, although I worried about what was going on.

The next morning around six o'clock, Sheriff George Wilcox, a good friend of my father's, went to my folks' house with one of his deputies. George said he thought I was in a lot of trouble at the base. He told my dad, "You tell Glenn to keep his mouth shut, if he knows anything. They want all your kids' names, they want to know when they were born and where they are now."

Dad said Wilcox was really shaken up. My dad got in his car and came to my house, the funeral home, as fast as he could. He almost knocked the door down, and he got me out of bed. It wasn't much after 6:00 A.M. I got up and Dad and I went outside, and I finally told him what happened, just like it happened. At first he said our government wouldn't do a thing like that, then he got to thinking about it. He said I'd never lied to him, so it must be true. Then he got very angry, but he said he wouldn't talk about it because he didn't want me to get killed.

Later that morning the nurse called me, crying. She said she had to talk to me. I suggested the officer's club, which was only about a block from her quarters. She agreed, and I drove straight out there. She was almost hysterical, sick to her stomach, and ash white. She was in uniform, but really disheveled. She wanted to know what happened to me. I told her what they did to me, but that I didn't know why. She said, "Well, I'll tell you why!" She said she found out later that all the regular infirmary staff had been told not to report for duty. Somehow she didn't get the order.

She went to work as usual and went into the supply room to get her day's supplies. When she did, there were two men, doctors, in surgical masks and everything. There were two gurneys with unzipped body bags on each one. The doctors were at one gurney, with the bag folded back. There were two small mangled bodies in the bag. She said the smell was the most horrible she'd ever experienced. The doctors said something about it being toxic, but I can't say what that meant. She said they ordered her to come over and told her, "We need some help. Lieutenant, you have to take notes for us, write down what we're looking at, what we tell you." She wrote down everything they said as they examined the bodies.

I asked her if she knew the doctors, or where they were from. She said she had never seen them before. She told me she heard one say to the other that they'd have to do something when they got back to Walter Reed Hospital.

She then described the bodies to me. She said a hand was severed from one of the bodies, and they turned it over with a long forceps. There were only four fingers. They had little pads on the tips with what looked like tiny suction cups. Their mouths were only slits, about an inch wide. There were no teeth, only a firm piece of tissue-like cartilage. They had no earlobes. The nose was concave, with two orifices, but no bridge. The eyes were very large and sunken so far back in you couldn't tell what they looked like. She said the heads were disproportionately large, and the doctors noted the skull structure was like a newborn baby's.

Some people think this nurse never existed. That's because I never mention her name. I gave her a sacred oath that I wouldn't. That's what I called it, a sacred oath.

Walter Haut

Lt. Walter Haut was the Public Information Officer for the 509th Bomber Squadron. He wrote the press release about the retrieval of the Roswell crash debris.

Col. Blanchard called me to his office to issue a press release. I'll never forget it, and I'm beginning to think no one else will, either.

Lt. Walter Haut

"I want a press release on the stuff the rancher brought in. You don't have to get too detailed. Just that he brought in a flying saucer, and that the base is not in possession of it. Fill in the rest and hand-deliver it to the media."

It was morning, so I spent the rest of the time writing the release. It was already too late for the morning paper, but I made the deadline for the Roswell *Daily Record.* Radio could use it at any time. After that I went home and had lunch. It was about two o'clock when I returned to the base. The phone was ringing off the hook. The story had gotten on the wire service, and we were getting a lot of inquiries.

I returned home about 5:30 or so. One of two things probably happened. Either the phone quit ringing for a short period of time, or I got the feeling I couldn't answer the same question again and just left. I probably got some of the referrals from Sheriff Wilcox. The others he referred must have gone to headquarters. At the time, I didn't check into details like this. The next day they said it was a weather balloon. At least until ten years ago, that was the end of the story.

People started doing some research on it, primarily Stan Friedman, and Bill Moore and Charles Berlitz putting out *The Roswell Incident.* That probably gave the biggest kick to it. Prior to that there had been an occasional curious individual or a rare representative of some media.

You would think I would know more about what happened, since Jesse Marcel lived only a block from me. My wife and I played bridge with the Marcels from time to time, and occasionally we would visit. Remember, back in 1947 there were still visible barriers. He was a major, field-grade officer; I was a lieutenant, company-grade officer. Although we were pretty good neighbors, we were never real close.

Because I'm the one who issued the press release, there has been a lot of attention directed at me. Some think I know more than I do, and nothing I do or say will change their minds. The added publicity these last few years has been something of a strain, and I do hope the government comes out with the truth.

Frank Kaufman ("Steve Arnold")

Frank Kaufman, a sergeant in the CIC detachment at the Roswell Army Air Field, says he was on duty the night base radars tracked the UFO that eventually crashed outside of Roswell. In order to protect his identity, Kaufman originally called himself *Steve Arnold* in interviews.

But it was Frank Kaufman, in his own name, who told *UFO Magazine* that he was detailed by the Army to stay in Roswell after the war as a disinformation specialist to keep people from learning the truth about the Roswell crash. After the story broke and Jesse Marcel went public, Kaufman felt it was time to tell his tale as well.

Some have questioned Kaufman's credibility, but he paints an interesting picture, nonetheless. His story begins the night of the crash.

> The radars weren't working right. At least that's what we thought at first. After it was confirmed they were in perfect working order, we had to accept that something was flying around the country we didn't know about. My job was to help determine what it was. Man, was I going to be surprised.
>
> Maj. Robert Thomas, my boss, had just flown in from Washington and led our group to White Sands to monitor the situation. We would see for ourselves what these erratic radar blips were all about. My superiors back East were concerned. Gen. Scanlin also called about the problem. Radar wasn't perfected then as it is today. The blips didn't stay on the screen long. They would fade out until the radar antennae again passed the target and confirmed its presence. Watching a radar screen can be boring, but we kept our eyes on the screen as much as possible.
>
> Occasionally, we went to another room. A radar operator put a mirror on a chair, and if something unusual happened he bounced a light off of it into our room to alert us.

Frank Kaufman

On July 3, radar screens still showed erratic movements, pulsating several times. Then a blip exploded into a starburst on the screen. Normally a plane or missile wouldn't do that. We thought something had gone down, but didn't quite know where; somewhere to the east of White Sands, behind White Mountain. Radar couldn't see behind that mountain range, and whatever it was that went down could have descended just about anywhere.

Roswell was generally in the correct impact direction, and Thomas decided we should return. We needed to get our heads together and figure out how to proceed from there. En route, we were in constant radio contact with the base.

Our best clue came from a motorist traveling to Roswell from Vaughn on U.S. 285. He noticed a flaming ball falling to earth and called the base. Others in the area also noticed something fall. It was common practice at that time for people to report strange or unusual aerial sightings to the base.

A military detail was immediately sent out to locate the area. We organized our own detail to include heavier equipment that would follow as soon as possible. The scouts located the area about thirty miles north from the base gate and corralled some interested civilians before we arrived. Access to the area was generally secured by the time we caught up with the advance detail.

We were guided by an orange glow or aura above the craft, a hazy light that could be seen for some distance. After clipping a few fence wires, we finally got to the site. The terrain there was generally flat except for the hill the craft impacted into. About a hundred yards from the impact site, we stopped. Vehicles formed a semicircle around the craft.

We were transfixed by what we saw. The bottom of the craft was grayish white. A soft white light came from cell-like structures that covered the bottom of the craft. That, coupled with the orange aura above the craft, formed a bizarre picture. The craft was sort of triangular in shape, with what we supposed was the front end crumpled, and the rest embedded into the ground at an angle of about thirty degrees. From wing tip to wing tip, it was about twelve feet; front to back about twenty-five feet, more or less. There was an opening on the left side where a seam had split open.

Searchlights illuminated the scene, and men in chemical suits advanced first. They brought all kinds of instruments to the object. There were some readings, we were told, but not of what. The orange halo around the craft began to dissipate. Photographers approached next to record it all before anything was moved.

We noticed two bodies near the craft; one was partially out of the damaged opening on the side. As I remember it, they were somewhere around five feet in height and well proportioned for people that size. They had fingers like ours, but their heads and eyes were larger, although not a lot.

Their skin appeared human-like to me; ashen in color like dead people. We noticed deterioration setting in, so we immediately put all the bodies in shiny silver-colored body bags, then placed them on a couple of Jeeps to be taken to the highway, transferred to a hospital truck, and transported back to the base. I didn't go with that detail.

After that, I got a chance to briefly look into the cabin of the craft. A person five feet tall could stand inside comfortably. I noticed a series of panel controls and flickering lights. I remember seeing some unusual symbols, something like an *S*, *J*, and *O*.

We brought a large truck with a crane, and guys from the motor pool rigged up something to lift the craft. I don't know if it was heavy or not, but the truck moved very slowly. All of the stuff was taken to Hangar 84.

Then we made people line up shoulder to shoulder and pick up all the little pieces. We made more than one sweep of the area, and later we brought a large commercial vacuum from the base to make sure we didn't miss anything.

While at the site, we decided diversionary areas were needed since several people had reported seeing something crash. We brought a bunch of balloon pieces and aluminum scrap from the base to be taken to areas away from the hot spot. This was all done before daybreak.

We had been up for over twenty-four hours and there was still a lot to do. We orchestrated a press release. This was part of the cover-up. More pictures had to be made of the stuff in the hangar. At night and with all the lights out, we loaded planes to transport the material. Some planes were decoys, some weren't. They had various destinations; Andrews Air Base was one. I suppose some big politicians wanted to see the stuff. Other destinations were Fort Worth and Wright-Patterson. Later, we concentrated on debriefing the people involved in small, manageable groups.

All of this was over fifty years ago, but I have never forgotten the faces of the aliens; so peaceful.

Phyllis McGuire

Phyllis McGuire's father was Chaves County Sheriff George Wilcox, one of the central witnesses to the Roswell incident because he took the box of debris from Mac Brazel and put it in a jail cell, then called the 509th and spoke to Jesse Marcel. Phyllis was in Roswell the night of the crash, visiting her father. Years later, after her father's death, her mother confirmed what happened that night.

Phyllis McGuire

The radio and newspaper stories had played up this [Roswell] stuff for several weeks previously, and now it was a sensation. My desire to know more demanded instant satisfaction.

My father's job had the living quarters and office in the same building, the county jail. I walked into my dad's office; he was sitting at the big roll-top desk. No one else was there; it wasn't usually a busy office anyway. "What about this flying saucer I've just read about in the paper?" I asked. "Is it really true?"

"I don't know why Brazel would have brought that box of stuff in here if there wasn't something to it," he said, nonchalantly. "Some of it looks like tinfoil, but isn't. When you wad it up and then open your hand, it just comes right back out flat.

"I sent some guys out there, but it got too dark to see very much. They said they saw a big black circle of burnt grass about the size of a football field, but because it was dark, they turned around and came back. I sent them back out there the next morning, but the soldiers were already there and had it blocked off. Nobody was allowed to get to the area."

"Why did you call the Army in the first place?" I asked, irked at what the Army had done.

"We've had an agreement with the Army for a long time to call them about things that affect the military. It's their business, and they have more personnel and equipment to handle problems than we do. That goes for crashes, too."

The paper called it a flying saucer, and my father confirmed that it was unusual. People from all over the world called him on the telephone. He was up all night just talking on the phone. When he finally returned to the living quarters, he said, "I just got through talking to London, England."

That was unusual then, because we didn't talk overseas that easily. Things were different. There were no television stations to call that could send helicopters to take pictures for us to see immediately. We didn't have instant access to the news. That's why this thing could be covered up. I don't think I even knew the term *cover-up*. At least it wasn't part of my vocabulary back then.

After those initial talks with my father, I kept going back and he'd say, "I don't know anything—nothing has happened. I don't know what else is going on." He told my mother he wished that I would quit asking him questions. I think he felt badly about not telling me something. Then again, he might have been trying to protect us.

Several days later, on a Saturday morning, the Army descended on us again. Suddenly, three Jeeps and two cars all painted olive drab with big white stars on them drove up to the office. Although I didn't keep watching them, I think they stayed quite awhile. They came back to pick up all the material that was stored in a room at the sheriff's office. I don't know what it was in there.

My father didn't want to talk about it anymore after that, and asked my mother to tell me not to ask any more questions, so I didn't. Mother added, "That Air Force officer was really mean to George."

"Well, is it his fault the flying saucer came here?" I replied. I just didn't get it then.

I don't dwell on what happened in Roswell a great deal, and I haven't seen what people call UFOs, although I have met some credible people who have. Some people think that this just could not be, that it's not possible. I thought it was possible then and I think so now.

My mother and father were very close and nothing happened to one that the other didn't know about. In the early 1970s, she started telling me more about what happened.

"What was it?" I once asked her.

"Alien," she replied.

I knew it. I've always known it. Not consciously, but subconsciously. The truth brought instant relief to what had been disturbing me for so long. "Were there bodies?"

"Yes, there were three bodies. One of them was still alive when they were found, but it eventually died."

They had large heads and eyes but small bodies. She said my dad felt very sorry for them. I got the impression they weren't given good care. They were treated as enemies. She didn't talk about the material.

My dad told her the whole thing. She always knew. She didn't mention anyone else who was involved. She felt a little guilty about talking, because they were told not to say anything or the family would suffer the consequences. They said they would kill the whole family. I've heard others were threatened the same way.

Mother put a short description about what happened in 1947 on paper, which I suspect she wrote before talking to us. Possibly she was still afraid to talk, but was even more concerned that the story of what happened here would be lost.

Because of all the publicity on the subject the last few years, if anything is mentioned about UFOs, I'm forced to be defensive. There are still so many people who think that none of this happened. Conversely, many people who are interested in this field seem to be nuts. It puts us in an awkward situation. We've had relatives who thought we were a little bit strange. Then we have others who back us 100 percent.

Frankie Rowe

Frankie's father was a firefighter in the Roswell Fire Department, one of the first official units to reach the scene after the crash. When she tells the story of the bullying Army Air Force official who towered over her when she was a young child, scaring her into silence with threats because her father had been exposed to crash evidence, Frankie Rowe's eyes tear up. Hesitant still, Rowe shares a memory that again reinforces the issues that make the Roswell mystery news, even after all these years.

Frankie Rowe

I don't think anybody will ever really understand how completely terrified you can be of an event. We were not afraid of the spaceship that crashed; we were afraid of our own government. Some people today find that hard to believe, but it really happened.

I was twelve years old. My father had just returned home from his shift at the Roswell Fire Department. He was normally very calm and laid back, but this time he was real excited.

"I've got to talk to everybody. How soon is supper going to be ready?"

"It's going to be a few minutes. Why?" Mother replied. She and I were in the kitchen cooking while Donny and Pat and Suzie played outside. Jean was off at work.

"Well, hurry up, because I want everybody in here. I've got something really important I want to tell everybody." He was so excited he would stand on one foot, then the other one, hopping back and forth.

Mother and I hurriedly put food on the table, and we all sat.

"We got a fire call this morning," Dad said. "Something about a crash about thirty miles north, close to Black Water Draw. I thought it was going to be a range fire caused by a crashed plane. When we got there, were we ever surprised. We all got off the truck and just stood there, not believing what we were seeing. It was a flying saucer. Two of the crew were dead, but one was still alive."

"What did it look like?" Mother asked, not totally accepting what she heard.

"They're little people. Little people are in these things. The two that were dead had been placed in body bags. The one still alive was walking around. It talked to us in our heads without saying anything, but we all heard the same thing. It said it was sad over the loss of its comrades, and that no one could help. It told us they were not here to hurt us."

Father went on. "About this time the military arrived. They sure were angry we were there; chased us out real quick. We weren't even allowed to fix the fences we had to cut to get there. Sure hope no cows get out. They can be dangerous on the highway."

Mother again asked what the creatures looked like.

"They're only as big as a small ten-year-old child." I was 12, but only maybe four feet tall. I think Dad meant these people were really small.

"What do their faces look like?"

"They look exactly like the Child of the Earth." This was the only reference he ever used as to what they looked like. The "Child of the Earth" is an insect that looks like a baby. He never said anything about the color except that it was the same as the Child of the Earth, so we assumed it was pale pinkish gray.

"Is the body like the Child of the Earth?" Mother asked.

"No," my father answered. "The body is like a little person's. They wear clothes just like we do."

Roswell firehouse, photographed in 1997.

At one point in the conversation, Mother made fun of his story. He looked at her in dead seriousness and said, "What makes you so smug to think that a God that can create this earth couldn't create whoever he wanted to?"

A few days later, I was visiting the fire station. I was munching on a cup of ice when a state policeman walked in the back door. "Hey, Dan, call your guys over here. I've got something I want you to see," he said. I was the only one at the table when he walked over and stood next to me.

"I got away with something and I want to show it to as many people as I can because I don't know how long I'm going to get to keep this," he said. His right hand was in his pants pocket while he was talking. He pulled his fist out, held it over the table, face up. Then he turned his hand over and opened it.

Whatever he had was wadded up in his hand, but when it hit the table it was as smooth as glass. It unfolded immediately. He picked it up and dropped it again and again. "I can't cut it or burn it," he said.

So the firemen took out their pocket knives. One would hold it while another would try to cut. They tried to burn it, too. The edges were jagged, so it had been torn somehow. They left it on the table without picking it up, so I played with it for a few minutes, long enough to know it wasn't something I had ever seen before, or since.

I'm not sure what the time span was between that day and the afternoon when I looked down the road and saw a military car coming to our house. The car stopped, an MP got out and opened the door for another guy, and accompanied him to the door. One stayed at the car. The MP accompanying the man to the door eventually went to the back yard to stay with Donny and Pat and Suzie. When they got to the door, I told Mother.

She opened the door. The man wasn't dressed as normal military police. There was no black armband with the white letters "MP." He was in a fancy dress-type uniform, with the dress coat going over the trousers. It was olive drab. He had a belt that went around the waist and up over the shoulder, too. His eyes were covered by dark sunglasses, which he never removed.

"I want to talk to the person who was at the fire station when the state policeman was there," he demanded.

"That would have to be Frankie," Mother said. "She was there."

"Is that you?" He asked, looking at me.

"Yes," I admitted.

"We want to talk to you." Mother invited him in.

"Is there anyone else here?"

"Yes, the other kids are playing in the back yard."

He wanted to know how many there were and what their names and ages were. Mother told him and asked him to have a seat in the living room.

"No, I want to sit there at the table." He pointed to the dining room, just beyond the living room. We started to sit, but he said to Mother, "You sit at the end, I want her over there." So I was at the short side across the table from him, with Mother at my left. Refusing a chair, he stood facing me. Every time I relive this event, uncontrollable tears fill my eyes. I cannot do it without seeing and hearing this despicable man.

"What did you see at the fire station?" He demanded.

"I saw a piece of material that came from the flying saucer that crashed," I replied.

"No, you didn't. You didn't see anything."

At my age, I didn't understand his technique, so I kept insisting I saw it—and he kept insisting I didn't. He got very angry with me and raised his voice, something he didn't need to do. He had a voice that just boomed; it was very cutting and sharp. His accent was back East, New York or New Jersey.

He kept saying, "You don't understand what you have seen. You don't know what it was. You didn't really see anything." He took a stick-thing from his belt, and started beating his hand to emphasize the point. He was not necessarily a tall man, but he was broad, and standing over me he was huge.

"If you ever talk about this the rest of your life ... " He paused for several seconds, then started again. "You know, there's more than one thing that we can do to shut you up. First of all, we can kill you and every single member of your family. We'll take you out in the middle of this big desert here and we'll bury your bodies. No one will ever find them. Or we could take you to a prisoner-of-war camp; you children in one prison and your parents in another."

He looked at Mother, "Can she keep a secret?"

"If you tell her she needs to, she can," Mother said.

"You have to make a promise to me that you will never talk about this, about what you think you saw, because you did not see anything. Nothing about what you think you handled, because you did not handle anything. You can never talk about this the rest of your life."

After I made the promise, I was crying, and Mother sent me to the bedroom. She never said much during that session but "Yes, Sir" and "No, Sir." I didn't understand that and was hurt that she didn't come more to my defense.

I kept the promise until they started doing all the research in Roswell in the early 1990s. Kevin Randle called several times. I met with him and was surprised that he was able to supply about 99 percent of the information. But as he talked, I realized he had to be talking about a different event. The one I knew about was closer to Roswell. When I mentioned it to him, he looked at me like I had two heads. Later, I think, he accepted that viewpoint. He told me that other people had handled some of the material also. It certainly would have been nice to know earlier that I wasn't hallucinating.

This has been a hard thing to go through. To this day when I hear a voice that sounds like the man that put me through so much hell, I start crying. Randle set up a phone interview at Walter Haut's house. When it came my turn to talk, I picked up the phone and immediately dropped it. I got hysterical and cried. All I could see in my mind was that guy beating his hand with that baton. They had to find an announcer with a different voice and accent. I suppose I'll be influenced by that inquisition for some time to come.

The Ines Wilcox Diary

Ines Wilcox

Before Roswell gained a prominence far beyond that of most small American towns, Sheriff Wilcox's wife, one of the few social reformers in the area, was involved in community affairs to an admirable degree. Fortunately for posterity, she kept a personal diary chronicling her earnest efforts to support her husband's goals of reform. This excerpt was provided exclusively to *UFO Magazine* by daughter Phyllis McGuire. In it, Mrs. Wilcox reflects the style of a close-knit community sincerely interested in the welfare of its residents. Sheriff Wilcox and his wife focused on law enforcement protocols and renovation of the jail itself, but it was the subject of her casual entry about the saucer crash that was to enthrall the nation's populace in years to come.

In her journal, *My Four Years in the County Jail,* Ines Wilcox describes the night Mac Brazel appeared at the Chaves County jail with a box of debris from the crashed spacecraft. She wrote this entry years before Jesse Marcel went public with his story and before Stanton Friedman and Kevin Randle wrote their respective books about Roswell. She made this entry privately, with no intention of ever writing a book or seeing her work made into a television movie. In other words, just from the circumstantial evidence alone, Mrs. Wilcox believed that what she was writing was the truth.

> One day a rancher north of town brought in what he called a "flying saucer." There had been many reports all over the country by people who claimed they had seen a flying saucer. The rumors had many variations. The saucer was from a different planet, and the people flying on it were looking us over. The Germans had invented this strange contraption, a formidable weapon. There were other tales, that one had landed and strange-looking people all seven feet tall or more walked from it, but quickly departed on sighting any onlooker.
>
> All the papers played the stories up, and many people searched the skies at night to catch sight of one. Since no one had seen a flying saucer, Mr. Wilcox called headquarters at the Air Force base and reported the find. Almost before he hung up the telephone,

an officer walked in. He quickly loaded the object into a truck, and that was the last glimpse anyone had of it. Simultaneously, the telephone began to ring; long-distance calls from newspapers in New York, England, France; government officials, military officials. The calls kept up for twenty-four hours straight. They would speak to no one but the sheriff. However, the officer who picked up the suspicious looking saucer admonished Mr. Wilcox to tell as little as possible about it and refer all calls to the Air Base. A secret well kept, for to this day we never found out if this was really a flying saucer.

China, UFOs in

In China, UFOs and UFO sightings are a matter of great national importance. A sudden proliferation of sightings began in December 1999, causing enough of a sensation to break into the normally tightly controlled national news media.

Since December 2, 1999, when people in Shanghai observed a shining cylinder with a flaming orange tail "scooting" above the skyscrapers, almost every city in China has had a sighting or two. In northern Changchun, three "suns" appeared in the sky, while in southwestern Chengdu, a V-shaped UFO was sighted. Dalian, Shantou, Xiamen, and other cities reported glowing streaks on the horizon, and Internet sites began posting blurred photographic evidence.

Still another sighting took place on December 11, 1999, in an impoverished farming community. According to Associated Press writer Charles Hutzler, "Poor farmers in Beijing's barren hills saw it: an object swathed in colored light that some say must have been a UFO."

China's largest UFO research organization is headed by Sun Shili, a professor of international trade at Beijing's University of International Business and Economics. Sun, who once worked as a translator for Chairman Mao, had a sighting of his own in 1969 while working on a "learn-from-the-peasants" campaign at a rural farm. "While planting rice," Sun told the *Christian Science Monitor*, "I saw a glowing sphere flying oval-shaped orbits between the ground and the sky, but at first I thought it must have been a Soviet spy ship." It was only after Sun received a Spanish UFO book to translate that he realized what he had seen might not be from this world. When Mao died in 1976 and a new spirit of candor on the subject began to take hold, Sun continued to do more research on the phenomenon, quickly becoming China's leading UFO expert.

Along with a more open interest in UFOs, China is also in the beginning stages of creating its own SETI program. The academy is awaiting funding to begin the construction of a radio telescope, even as Chinese newspapers and broadcasters, including the staunchly traditional, government-operated Chinese Central Television, are disseminating more and more reports of close encounters with aliens.

Chinese UFO Research Organization (CURO)

China's largest UFO research group, a branch of the Chinese Academy of Social Sciences, that has over 20,000 members. *See* **China, UFOs in**

Chorley, Dave

The other half of "Doug and Dave," the English crop circle hoaxers. *See* **Bower, Doug; Crop Circles**

Chorwon, Korea

In the skies over Chorwon on May 31, 1952, during the Korean War, military ground forces directed fire at an apparent alien spacecraft. The craft returned fire by emitting a pulsating ray that was witnessed by military pilots and ground forces. The men on the ground who were hit by the ray became so weakened they could not walk.

"Christine"

This is a "Men in Black" case. "Christine" was a Navy pilot stationed at a military base that was buzzed by a UFO. In a private interview, Christine said:

> One night a large UFO showed up and hovered over the base for about forty minutes. Everybody saw it. We were all pilots and we knew it wasn't ours. The brass didn't know what it was either. When it left, it took off at a phenomenal speed.
>
> The next day a bunch of black sedans showed up on the base. Each one had four guys dressed in black. They got out and went to every building on the base and "encouraged" us not to talk about what we saw. I'm not easily intimidated. I know how to use a firearm and I've flown combat missions. But these guys meant business.

Churchill, Sir Winston

The Prime Minister of the United Kingdom during World War II and again in the 1950s. Churchill once wrote a memo to Air Lord

Cherwell, asking him whether flying saucers were real. Lord Cherwell assured Churchill that they were a natural phenomenon. *See* **Cherwell, Lord**

CIA

See **Central Intelligence Agency**

CIC

See **Counter Intelligence Corps**

Citizens Against UFO Secrecy (CAUS)

Action group founded by attorney Peter Gersten. *See* **Gersten, Peter**

Clark, Alan D.

Base commander of Fort Worth Army Air Field. Col. Clark took custody of a special courier pouch upon its arrival at the base, a pouch sent by Col. Blanchard of the 509th Bomber Squadron containing debris from the Roswell crash. Col. Clark handcuffed the pouch to his wrist and boarded a B-26 for a flight to Washington, D.C., where he and his cargo were received by Army Maj. Gen. McMullen. As UFO researchers have pointed out, this was an especially high level of security to maintain for some simple weather-balloon samples.

Clementine I

NASA's 1994 unmanned mapping probe of the moon, launched at the encouragement of the Office of Strategic Defense Initiative. Clementine I discovered that the variation in lunar topography is much greater than NASA scientists and astronomers had believed and that the tallest mountains on the moon are higher than earth's Mount Everest. The probe also detected hints of water vapor at the moon's poles.

The Clementine mission has raised several warning flags to UFO researchers: Why have many of the photographs been locked away from public view, and why would the organization responsible for "Strategic Defense" be interested in the moon in the first place? *See* **Strategic Defense Initiative (SDI)**

Clinton, William Jefferson

Shortly after his inauguration as forty-second president of the United States, Bill Clinton asked his number-two man at the Justice

Department, Webb Hubbell, to find out who killed President Kennedy and what the truth is about UFOs. Shortly after Hubbell began his investigation, inquiries into "Whitewater" and the allegations by Paula Jones began. There are members of the Clinton administration who, to this day, maintain that Clinton's troubles had more to do with his desire to open up the nation's UFO files than with anything else. Hubbell claimed he was "stonewalled" by the people he sought answers from, and was forced out of office and indicted by Special Prosecutor Kenneth Starr.

On a trip to Ireland, however, President Clinton singled out one letter he had received from a Belfast youngster named Ryan. He told the child that, "No, as far as I know, an alien spacecraft did not crash in Roswell, New Mexico, in 1947." Clinton's response aroused some laughter from the crowd. "And, Ryan," the president added, "if the United States Air Force did recover alien bodies, they didn't tell me about it, either, and I want to know."

However, President Clinton also signed an executive order maintaining secrecy over Area 51, in the face of a claim by base workers that they were being exposed to toxic chemicals in violation of EPA regulations. Clinton's order exempted Area 51 from releasing environmental reports to the public. The veil of secrecy was extended in September 2003 for an additional year. *See* **Hubbell, Webster**

Clonaid

Affiliated with the UFO cult the "Raelians," Clonaid announced that its first cloned child had been born in 2002. Clonaid's program has been questioned on many grounds; the organization received a large fee for its services, the debate over the morality of cloning is far from being settled, and the organization's medical procedures are controversial. Cult members believe humanity has been cloned from alien ancestors, and its philosophy insists that the practice will someday become common and eventually lead to a form of immortality.

According to program head Dr. Brigitte Boisselier, the cloned baby's parents had been unwilling to identify themselves and provide DNA evidence to substantiate their claim for fear of losing legal custody of the child. A Florida lawsuit attempted to accomplish just that but was eventually thrown out of court. Soon after the judgment Clonaid announced the births of two other cloned children, one in Europe and another in Japan. *See* **Boisselier, Brigitte; Rael; Raelians**

Close Encounter Research Organization (CERO)

Founded and managed by abduction therapist Yvonne Smith, this organization attempts to provide a safe haven for those individuals who have undergone especially traumatic alien abduction or contactee experiences. *See* **Smith, Yvonne**

Coleman, William T.

Air Force major who disclosed a 1962 order directing the Air Force Office of Information to "delete all evidence of UFO reality and intelligent control," which contradicts the Air Force stand that UFOs do not exist. The same directive applies to AF press releases and UFO information given to Congress and the public. The Air Force is on record ordering its own information service to withhold information on a potential national security threat from Congressional oversight. *See* **Air Force, Studies of UFOs**

Collective Unconscious

A term coined by psychologist Carl Jung to describe a shared pool of information and imagery common to the human species. It has been said that remote viewers, as well as shamans or "spirit walkers"—those who have mastered the technique of astral projection or out-of-body travel—are able to leave this dimension of reality and dip into the pool of the collective unconscious, where they can move about freely. In his book *Flying Saucers: A Modern Myth of Things Seen in the Sky*, Dr. Jung suggests that flying saucers are not actually real extraterrestrial spacecraft but a shared human image.

Collins, Paul

One of the witnesses to the "Los Angeles Air Raid" on February 25, 1942, which the U.S. Army attributed to a Japanese attack. Some have claimed the "Raid" was actually an overflight of a formation of flying saucers. The objects, whatever they were, traveled over many parts of Southern California, successfully eluding intense anti-aircraft fire. Collins, a Douglas Aircraft employee, estimated the speed of the objects at five miles a second *See* **Los Angeles Air Raid of 1942**

"Colonel Rogers"

In another reported "Men in Black" case, "Col. Rogers," a pseudonym for a retired military officer, told friends that he had once seen a UFO while in the service and that they were real, not exotic military weapons. Several days after relating his story, a man dressed completely in black and wearing sunglasses showed up at Col. Rogers' front door. The stranger warned Rogers to keep his mouth shut regarding UFOs. Rogers then brandished a loaded .45 pistol and pointed the weapon at the man's head, telling him, "I don't know if you're CIA, NSA, or an alien, but somebody will have a lot of explaining to do if I blow your head off. Now get out of here."

The next day another man showed up wearing regular business attire, hailed the colonel from the front yard with his arms raised, and told him, "Look, Colonel, we're sorry. No more guys will show up at your house. Some of this UFO information is still classified and we would appreciate it if you just didn't talk too much about it. Okay? Thanks." And then he left. *See* **Men in Black (MIB)**

Colorado, University of

Site of the deliberations and of the now famous Condon Report of 1968, which resulted in the closing down of Project Blue Book in 1969. See **Condon Report; Projects, Government UFO**

COMETA

See **French Flying Saucers Report**

Condon, Edward U.

In 1969 Dr. Edward Condon chaired a University of Colorado panel tasked with studying UFOs. Condon was a nuclear physicist who had worked on the Manhattan Project during World War II. He had survived an inquiry by the House Un-American Activities Committee in the 1950s and an attack by Richard Nixon himself before being exonerated. Allowed to pursue his career in academia afterwards, Condon became in every way a government "insider."

The Air Force's stated position was that UFOs did not exist and were therefore not a threat to the national security of the United States. But public and Congressional pressures had forced the service's hand, and they needed a formal "stamp of approval" to get out of the "saucer business."

At the Air Force's urging, Condon, now at the University of Colorado, took on the unenviable task of chairing a panel that would appear to conduct an independent examination of all the UFO evidence, but in effect rubber-stamp the Air Force's predetermined conclusion. Condon's own inclinations led him to believe it would be in everyone's best interests that the panel should "dispatch" the saucer theory as quickly as possible. His executive summary, fronting the panel's final report, was enough to provide the Air Force the excuse it was looking for to shut down its own Project Blue Book. However, Condon's preconceived conclusions were completely at variance with the rest of the University of Colorado Study, which determined that UFOs were a real phenomenon that needed further investigation.

Edward Condon, whose preface to the University of Colorado Report dismissing the possibility of flying saucers was a pretext for the Air Force cancellation of Project Blue Book.

One of the ironies stemming from this whitewash was Condon's own evaluation of UFO sightings made by American astronauts. Referring to one of the Gemini earth-orbital missions during which the astronauts saw a UFO, Condon wrote, "The training and perspicacity of the astronauts put their reports of sightings in the highest category of credibility. Especially puzzling is the Gemini 4 mission sighting of an object showing details such as arms protruding from an object having no noticeable angular extension." *See* **Condon Report**

Condon Report

One of the most controversial studies of UFOs and, some say, an attempt at disinformation at the instigation of the Air Force. At first, two major organizations, NICAP (headed by Donald Keyhoe) and APRO, welcomed this investigation and offered to open their files to the Condon Committee. However, Robert Low, assistant dean of the graduate school at the University of Colorado, who would be chosen to run the committee's day-to-day activities, wrote a letter to UC faculty officials insisting there were reasons to oppose such a study.

Low's problem with the entire project was that in order to study UFOs objectively, one had to approach them from the possibility that "such things exist." This was "not respectable"—a position that meshed completely with Dr. Condon's own notions. So Low concocted an approach that would focus on the credibility of witness testimony rather than upon the scientific basis of the saucer phenomenon itself.

This bit of academic sophistry allowed the project to proceed, but not without some major hiccups along the way. The thirty-six panel members who participated in the 1969 Condon Report found many cases that seemed to indicate that UFOs were of an extraterrestrial nature. But Condon and Low realized that the focus of public and media attention would fall on Condon's executive summary, not on the fine details. As predicted, the conclusions stressed by Condon overshadowed the rest of the panel's. With the study completed, the Air Force quietly shut down Project Blue Book and buried the UFO subject from public view once and for all. Despite outcries of whitewash, the Condon Report would remain the final official government verdict on UFOs until the Rockefeller Panel in 1996 all but reversed the Report's findings. *See* **Central Intelligence Agency, History of UFOs in; Keyhoe, Donald; National Investigations Committee on Aerial Phenomena (NICAP); Sturrock, Peter**

Contactees

These are individuals who claim to be in contact on an intellectual level with beings from other worlds or dimensions. Contactees are voluntary abductees, in a sense, or self-proclaimed prophets in another sense. George Adamski claimed to be a contactee, as did Swiss farmer Billy Meier, Betty Andreasson, and Philip Krapf. Contactees sometimes try to provide credibility to their stories by making predictions about the future and claiming these prophecies were provided to them by their alien brothers. *See* **Abductees; Adamski, George; Andreasson, Betty; Krapf, Philip, Meier, Eduard "Billy"**

Cooper, Gordon

One of the original Mercury astronauts, Col. Gordon Cooper was the last American to fly into space alone. In 1978 Col. Cooper appeared before the United Nations General Assembly and asked for "open discussions" on the UFO matter. Cooper said the secrecy started either during or shortly after World War II. He told the UN, "I think they thought the public would be frightened by knowing somebody had some vehicles that had so much better performance than anything we had, that maybe it would create panic if the public learned about them. They probably tried hushing it up until they found out more about them. Then as time went on, it just got more embarrassing trying to cover it up."

Cooper's professional career has included several UFO experiences. In 1978 he said, "I did have occasion in 1951 to have two days of observation of many flights of them, of different sizes, flying in fighter formation, generally from east to west over Europe. They were a higher altitude than we could reach with our jet fighters of that time."

And in a 1997 interview with the *National Enquirer,* Cooper was reported to have said:

> I had a camera crew filming the installation [of a precision landing system at Edwards Air Force Base in California] when they spotted a saucer. They filmed it as it flew overhead, then hovered, extended three legs as landing gear and slowly came down to land on a dry lake bed.
>
> These guys were all pro cameramen, so the picture quality was very good. The camera crew managed to get within twenty to thirty yards of it, filming all the time. It was a classic saucer, shiny silver and smooth, about thirty feet across. It was pretty clear it was an alien craft. As they approached closer, it took off.
>
> After a while, a high-ranking officer said that when the film was developed I was to put it in a pouch and send it to Washington. He didn't say anything about me not looking at the film. That's what I did when it came back from the lab and it was all there just like the camera crew reported.
>
> I had a good friend at Roswell, a fellow officer. He had to be careful about what he said. But, it sure wasn't a weather balloon like the Air Force cover story. He made it clear to me what crashed was a craft of alien origin, and members of the crew were recovered.

Cooper, Milton William

Milton William Cooper was shot and killed by police officers after he himself opened fire and wounded one of the officers while they were in the process of serving a felony warrant on him. Cooper was a conspiracy theorist who supported the idea that it was the Secret Service driver of JFK's limousine in the Dallas motorcade that turned and shot the president in the head. Cooper became involved in a number of Area 51 theories, later claimed to be a disinformation specialist "testing" the UFO community, and finally distanced himself from the UFO field, associating instead with the ultra-right-wing community before the federal government filed bank fraud charges against him. He is best known within the UFO community for a debate over the legitimacy of the MJ-12 documents and for his associations with the Hollywood community in early 1990. *See* **Area 51; Majestic**

Copley Woods

Abduction case examined by Budd Hopkins and made famous in his book, *Intruders: The Incredible Visitations at Copley Woods. See* **Hopkins, Budd**

Corona, New Mexico

Some UFO researchers suggest this site was the real location of the "Roswell" crash. Corona is actually miles closer to the location of the debris field than is the city of Roswell. However, the crash debris was taken to the Roswell Army Air Field, and Mac Brazel brought the debris he collected to the Chaves County sheriff's office in Roswell. Therefore the name *Roswell* has become attached to the event. *See* **Roswell, New Mexico**

Corso, Philip J.

A U.S. Army artillery officer, then an intelligence officer (G2), Lt. Col. Philip J. Corso, who died in 1998, retired from the service in 1962 as the deputy director of Army Research and Development, where he'd been, at one time, the director of the Foreign Technology Division.

Philip J. Corso (l) with Day After Roswell *co-author William J. Birnes in Roswell, New Mexico, 1997.*

In his book *The Day After Roswell*, Corso wrote that at Army Research and Development, he was responsible for taking the Army's remaining cache of Roswell technology—stored in a Pentagon basement file cabinet—and secretly slipping it to those defense contractors and industries working on the same technologies. The Pentagon Roswell file contained a handheld surgical laser device, a fiber optic junction, bits of solid-state circuitry, atomically-aligned fabric, and a headband with electromagnetic pickups that Corso said was worn by the alien pilots of the crashed Roswell craft. According to the medical examiners at Walter Reed Army Hospital and the Army's Research and Development technical advisors, the headpiece allowed the aliens to navigate their spacecraft without the use of flight controls, instead translating images from their brains into directional controls.

Before he arrived at Army Research and Development, Corso commanded a surface-to-air missile battalion on NATO's southern flank in Germany during the height of the Cold War, where he was one of a handful of frontline officers to have stockpiled nuclear warheads ready to use. During the Korean War, Corso was on Gen. Douglas MacArthur's staff, where he was responsible for developing a list of potential nuclear targets in North Korea, had MacArthur been given the authorization from President Truman to use nuclear weapons.

At the end of the war, Corso was the Army's representative for the exchange of prisoners of war. He warned then-President Eisenhower that the North Koreans were holding back American GIs and pilots, meanwhile using their identities for the insertion of deep-cover operatives into American society who would be called upon to perpetrate sabotage and other terrorist acts in the event of a war with the Soviet Union.

Corso's claims, although heavily disputed at the time, were later proven to be correct when the release of memoirs of former KGB

Left to right: Czech General Jan Sefina, defense contractor and author John Douglass, Lt. Cmdr. and former CIA Station Chief Chip Beck, and Lt. Col. Philip J. Corso (retired) at Congressional hearings on American POW/MIAs.

officers revealed just such a plan (which was ultimately abandoned as unmanageable). Presumably some of those deep covers are still alive and in place in American society.

After retiring from the Army, Corso became the Director of Government Internal Security for a Senate committee headed by Senator Strom Thurmond. From there, Corso went to work for Senator Richard Russell, a member of the Warren Commission. Corso, at Russell's direction, investigated the Warren Commission and reported to the senator that the Kennedy assassination was a conspiracy directed by the CIA, funded by American crime families, and carried out by Cuban nationalists. He reported that President Johnson was complicit with the conspiracy and, as an accessory after the fact, had convened the Warren Commission to cover up the entire affair.

Cortile, Linda

A pseudonym for Linda Napolitano, one of Budd Hopkins' abductees, made famous in the 1979 Brooklyn Bridge case. Cortile told the story of having been floated out of her Manhattan apartment and "beamed" aboard a mothership while he husband remained asleep. The abduction was seen by many credible witnessed that evening. *See* **Brooklyn Bridge Abduction Case**

Linda Cortile

Counter Intelligence Corps (CIC)

The Army Counter Intelligence Corps was one of the few Army commands authorized to operate within the U.S. for military purposes. By law, the American military is not allowed to perform any law-enforcement functions except in an emergency. Army CIC units were present at the Roswell crash site. *See* **Roswell, New Mexico**

Crick, Francis

Along with James Watson, the discoverer of DNA, Crick was the first scientist to explain how DNA functioned to replicate itself from a genetic blueprint provided inside the nucleus of a living cell.

Crick subscribed to the theory called *panspermia,* which explains how life spread through the universe. This theory presupposes that an ancient civilization of extraterrestrials on a relatively benign planet set out to colonize the universe. Crick believes this was the only way life could have developed on a young planet like earth, since the conditions were far too hostile to sustain the spontaneous generation of living organisms from a basic primordial soup. Thus, Crick essentially became an advocate of a theory of the "old ones," extraterrestrials who were responsible for the seeding of life through the universe.

Years later, another DNA scientist, Kary Mullis, would claim to have had an encounter with aliens. Mullis also won a Nobel prize for developing the PCR method for making exact duplicates of DNA.

Crop Circles

Elaborate designs formed in wheat and grain fields, which have been discovered in England, Australia, eastern and western Europe, Japan, Canada, and the United States. Within these mysterious circles, the stalks of grain have been laid flat, usually in a circular or flowing pattern. Eighty-five percent of all crop circles have appeared in England in a concentrated area about twenty miles from Stonehenge. However, they appear annually all over the world, even near heavily populated cities. Crop circles first began appearing in grain fields in England in 1975.

Gypsy Patch at Etchilhampton, Wiltshire, UK; double glyph crop circle formation.

In 1990 the designs of these circles became increasingly more complex. Crop circle formations are absolutely stunning in their design, intricacy, and precise measurements. Many of them are huge, covering the space of nearly a dozen football fields. When carefully measured, the geometrical designs are accurate to within an eighth of an inch, even if the formation is a thousand feet long. Some of the more complex designs are called pictograms.

The stalks of grain involved in crop circles all exhibit cellular changes and an electromagnetic resonance that extends from the center of the crop circle design outward to the unaffected area, diminishing in resonance as it extends. Crop circle stalks have expanded nodes, and have been bent up to a 90-degree angle with no breakage. Although many have tried, no one has discovered any earthly process that can duplicate these cellular changes.

Julia set crop circle at Milk Hill in Wiltshire, UK, which, it was alleged, sprung up overnight.

Crop circles have a strong magnetic field in and around them, which is measurable. Many also exhibit strange patterns of radioactivity in which the level at a given spot will fluctuate. According to the laws of physics, this is an impossibility since levels of radioactivity are thought to be constant.

True crop circles often have several layers of grain in the design. Often the bottom layers will flow in one direction while upper layers flow in the opposite direction. The stalks have even been found delicately braided together. Often, thin metallic disks about the size of a quarter are found in crop circles. When these disks were analyzed by metallurgy labs at the University of Michigan and MIT, they were found to be composed of a combination of titanium, silicone, and oxygen. Both laboratories concluded that no industrial match could be found on this planet. In addition, when the disks were touched by a metal object, such as tweezers or a pen, they immediately turned into a clear liquid. The labs theorized that the disks had somehow been electrically charged, with the charge maintaining the molecular structure of the disk. When touched with metal, the charge was grounded and dissipated, which allowed the "metal" to return to its true liquid form.

Crop circles have become associated with UFOs because they often appear after UFO sightings. There's the assumption that because no one on earth can duplicate the energy process that imposes these designs on the stalks of grain, the cause must be extraterrestrial. However, this is not necessarily true. Most crop circles in England appear near the ancient structure known as Stonehenge. Many of the formations accurately reproduce the dimensions and configurations of Stonehenge to within a few centimeters.

Based on mathematical relationships and the theories of Euclidean geometry, the measurements of various crop circles appear to demonstrate many universal laws. Their measurements have been shown to correspond exactly to:

- *Notes of the musical scale.*
- *Electrical circuitry.*
- *DNA strands and genetic relationships.*
- *Sacred geometry/hyperdimensional physics.*
- *Ancient symbols found only in such places as the pyramids of Egypt, Stonehenge, and Mayan and Aztec temples.*
- *Solar system and asteroid trajectories.*
- *Mathematical equations.*
- *Advanced geometry.*
- *Molecular structure.*

Crop circles sometimes appear to be "footprints" of an actual three-dimensional form, a picture of a slice of a three-dimensional object, much like a CAT scan. Perhaps as corroboration of the three-dimensional aspect of crop circle theory, flocks of birds approaching a crop circle often split ranks to avoid flying above the formation, just as if a solid object were in their way. The flock then re-forms after passing the circle. Also, people who stand inside a crop circle have often reported experiencing a dramatic rush of emotions and increased perceptual abilities.

Crop circles have been the subject of hoaxes. In fact, two hoaxers have said that all crop circles are man-made. Doug Bower and Dave Chorley, a pair of English farmers both over the age of sixty-five when they announced themselves, claimed they were responsible for making the circles by using wooden boards and string.

As crop circles began to receive more publicity, they attracted thousands of sightseers trampling across private crops and farmland. Farm owners also claimed that they were responsible for the designs in an effort to stop the trespassing and loss in revenue from damaged crops. However, when asked to demonstrate how they made the circles, the farmers refused to try to create a formation.

A number of eyewitnesses claim to have observed crop circles being created within a matter of seconds. There are actual videos showing balls of light appearing over grain fields where crop circles seem to be created almost instantly. Though videos can be tampered with by the use of clever computer-graphics enhancement, many of the formations are so complex that it would take literally an army of people many hours to create it by hand. This leads many crop-circle researchers to believe that some credence must be given to a non-human source of this phenomenon.

The UFO-origin theory is most appealing to some researchers because it accounts for the correlation between flying saucer sightings and crop circles. At least one researcher, Doug Ruby, has said that crop circles are actually schematics for the assembly of a flying saucer. In order to prove this, he assembled a number of crop circle designs, combined them, and created a diagram for a spaceship that looked a lot like a Billy Meier "beamship." *See* **Bower, Doug; Chorley, Dave; Ruby, Doug**

Crow, Duward

A lieutenant in the U.S. Air Force assigned to NASA, who was the recipient of a letter from Col. Charles Senn, chief of the Air Force community relations division in which Col. Senn wrote, "I sincerely hope that you are successful in preventing reopening of UFO investigations." *See* **National Aeronautics and Space Administration (NASA)**

CSETI

See **Center for the Study of Extraterrestrial Intelligence**

CUFOS

See **Center for UFO Studies**

CURO

See **Chinese UFO Research Organization**

Cydonia, Mars

Unremarkable otherwise because of its relatively flat and featureless terrain, the Cydonia region of Mars, located in the planet's northern hemisphere, nonetheless appears to contain an artificially constructed humanoid face, large pyramids, and other features. These structures, which are alleged to feature complex mathematical relationships similar to the pyramids and Sphinx in Egypt, were photographed by NASA as far back as 1976, and the space agency has gone to great pains to deny their existence. NASA has since provided what it calls photographic "evidence" that "The Face" is a natural eroded mesa, though critics continue to allege otherwise. *See* **Hoagland, Richard; Mars, Face on; National Aeronautics and Space Administration (NASA)**

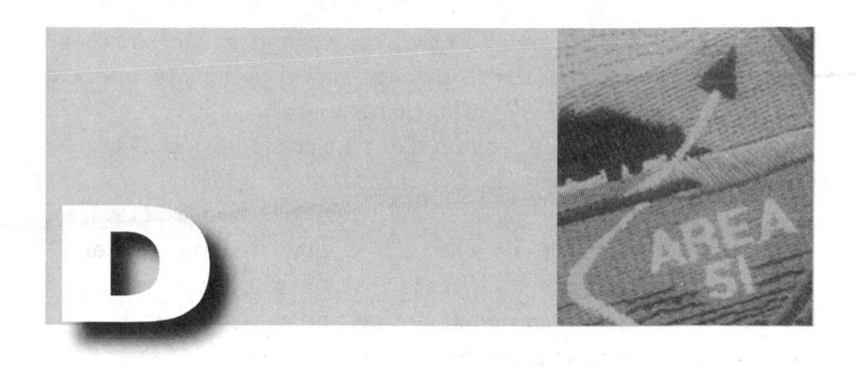

Däniken, Erich von

Author of *Chariots of the Gods?* and later works in which he theorized that earth had been visited in ancient times by extraterrestrials. In his book, von Däniken popularized the concept of "ancient astronauts" and explained many of the events of the Old Testament and other ancient writings as the intervention in human affairs by these extraterrestrials who manifested themselves as deities. *See* **Ancient Astronauts; Ezekiel; Valenov Plate**

Danzer, Roy

A local plumber in Roswell who was working at the base when the spacecraft debris was brought in and saw a dying alien on a stretcher near the base hospital. Danzer said that military officers warned him to stay silent—or else. See **Children of Roswell**

Darwinism

A naturalist theory in which biologist Charles Darwin sought to explain how life evolved from primitive to sophisticated forms through a process of natural selection and the survival of the fittest.

In the UFO community, proponents of an extraterrestrial source for life on earth have sought to disprove some of Darwin's essential theories by asserting that a simple analysis of the DNA of evolving species shows that human beings could not have evolved from earlier primates because apes and chimps have more efficient genetic structures than do humans.

Instead, looking at hominid DNA shows that humans are probably not native to planet Earth. This was part of the theory espoused by Claude Vorilhon ("Rael"), who claims that aliens he encountered in France in 1973 revealed to him that they created living species on earth by manipulating strains of earthly DNA.

In another theory, proponents of what is known as *intelligent design* insist that amidst chaos, there is an obvious, structured organization manifested in the higher forms of living things that could not possibly have been created as a result of that chaos. Just who the designer was is a matter debated by biologists, UFO believers, and theologians, but, according to mathematical models put forth by biologist William Dembski from Baylor University, intelligent design results from intelligent causation. *See* **Dembski, William; Pye, Lloyd; Rael; Raelians; Sitchin, Zecharia**

Daudpota, Azim

The Air Marshal of Zimbabwe, quoted in the *Times of London* on August 3, 1985 describing a UFO that was witnessed by many spectators. "This was no ordinary UFO," the air marshal said. "Scores of people saw it. It was no illusion, no deception, no imagination."

Davenport, Peter

Director of the National UFO Reporting Center, Peter Davenport runs a centralized data-collection agency that takes web-based and telephone reports of UFO sightings and reports them to local authorities, as well as to other agencies for investigation. His organization has become so well known that many 911 dispatch facilities actually direct reports of UFO sightings to Davenport. *See* **National UFO Reporting Center (NUFORC)**

Peter Davenport

Davis, Kathie

The pseudonym of one of Budd Hopkins experiencers in the Copley Woods abductions, part of a group of human subjects who, it was revealed, was part of a family of abductees and who had been a subject of long-term alien experimentation. *See* **Abductees; Copley Woods; Hopkins, Budd**

Dayton, Ohio

Home of Wright-Patterson Air Force base. In 1947 it was the headquarters of the Air Materiel Command where the UFO debris was taken after the crash at Roswell. There was also an important sighting of a UFO by a commercial pilot over Wright-Patterson, a sighting confirmed by radar and the visual sighting of Air Force fighters scrambled to intercept it, which they were unable to do. *See* **Roswell, New Mexico; Wright-Patterson Air Base**

Dean, Robert O.

Robert O. Dean

A retired Army sergeant major whose discovery of a 1964 NATO UFO report, which profiled four known types of ET including one that was almost human-looking, convinced him that not only were UFOs a real phenomenon, but that the governments of the world knew about them and were keeping information about them from the public.

Dean lectures UFO groups on the "who" and "why" of the ET presence, the ET origins of humanity, and the agenda of the government in opposing any disclosure of its relationship with extraterrestrials.

Debris Field, Roswell

The debris field was located at the Foster Ranch, approximately seventy-five miles outside of the city of Roswell. Among the witnesses who actually handled debris from this site were Billy Brazel, his father Mac, his sister Bessie, and Loretta Proctor (the mother of William Proctor) who accompanied Mac Brazel on horseback to the debris field the morning after the crash. *See* **Children of Roswell; Roswell, New Mexico**

De Brouwer, Wilfred

In 1991's *UFO Wave over Belgium, An Extraordinary Dossier,* Deputy Chief of the Royal Belgian Air Force, Maj. Gen. Wilfred De Brouwer, wrote:

> In any case, the Air Force has arrived at the conclusion that a certain number of anomalous phenomena have been produced within Belgian airspace. The numerous testimonies of ground

> observations compiled in this book (SOBEPS), reinforced by the reports of the night of March 30–31, 1990 have led us to face the hypothesis that a certain number of unauthorized aerial activities have taken place.
>
> The day will come undoubtedly when the phenomenon will be observed with technological means of detection and collection that won't leave a single doubt about its origin. This should lift a part of the veil that has covered the [UFO] mystery for a long time. A mystery that continues to the present. But it exists, it is real, and that in itself is an important conclusion.

Debunkers

A self-referential description and one also used within the UFO community, of individuals whose profession seems to be to discredit the existence of UFOs and those who witness such phenomena. Some debunkers attack all claims of the paranormal, asserting that it is merely the result of an illusion perpetrated by either a hoaxer or charlatan; some have even insisted they can fake paranormal events so as to prove how easy it is to stage such an "act."

Chip Beck's cartoon rendition of the missing Mars probe.

Within the UFO community, debunkers have sought to smear the reputations of witnesses, abductees, and researchers. Some debunkers are thought to be on the payroll of government agencies, as revealed in the Haines Report, and to discourage anyone, usually with ridicule, from coming forward to report or describe a UFO encounter. *See* **Klass, Philip**

Defense Support Program (DSP)

A series of surveillance satellites circling in high earth orbit, used to detect missile launches. Defense Support Program sensor platforms have also surveilled numerous UFOs, codenamed *fastwalkers,* entering the earth's atmosphere. DSP workers claim they have tracked thousands of such objects whose trajectories resemble that of intelligently guided craft rather than natural objects such as meteors. *See* **Fastwalkers**

Del Satto Observatory, Chile

In July, 1947 during the wave of international sightings, the observatory in Del Satto, Chile, records that they saw a formation of flying discs traveling at a speed of over 3000 miles an hour.

Department of Defense (DoD)

Known over the years as simply the Pentagon, DoD is the civilian command center for the American military, and as such is charged with protecting American national security and projecting American military might around the world as U.S. government policy dictates. It has long been interested in the subject of UFOs because flying saucers might pose a threat to the country. DoD is believed to use many agencies within its purview to actively study and contain the so-called alien "threat."

The Defense Intelligence Agency (DIA), founded in 1961, for example, acts as a repository for UFO reports from the intelligence services of foreign military agencies. The DIA has its own "spooks" and spying protocols and is the military's own spy agency, and it is often at odds with the CIA. *See* **Central Intelligence Agency, History of UFOs in**

Dembski, William

Biologist at Baylor University and proponent of the theory of "intelligent design," which espouses that life is not so much a product of nature as it is intelligently guided. Dembski's theory flies in the face of scientific naturalism, which asserts that "all phenomena are derived from natural causes and can be explained by scientific laws without reference to a plan or purpose." Dembski has developed mathematical models to detect intelligent design in nature, and if they hold up, the next problem will be to find the designer. *See* **Darwinism**

Dennett, Preston

Author of *Alien Healings*, which documents over a hundred cases of abductees who have been healed of grave illnesses by aliens using superior medical technology. These abductees were already the subjects of alien experiments, and Dennett theorizes that the aliens, far from being altruistic, were merely keeping their own medical subjects alive because they were part of a larger study and needed to be monitored. *See* **Abductees**

Dennis, Glenn

The local mortician at the Roswell Funeral Home in 1947. *See* **Children of Roswell; Roswell, New Mexico**

Díaz, Carlos

In 1975 Carlos Díaz, a businessman who lived in a remote part of Mexico, captured dramatic still and video photographs of hovering spacecraft. Spectranalysis of the film and videos show light frequencies never recorded before. Díaz, who claimed to have been abducted by humanoid creatures from the craft he photographed, so intrigued news correspondent and UFO researcher Jaime Maussan that Maussan provided him with a battery of cameras and film in a strategy to make sure that the resulting film could not be faked. *See* **Maussan, Jaime**

Disclosure Project

Initiated by Dr. Steven Greer as a way to confront the government face-to-face with the information it has been keeping from the American people about government knowledge of UFOs, ETs and the government's own dealing with these subject, the Disclosure Project has gradually grown in significance over the past few years as more and more members of the military, technological, and political communities have come forward with information about UFOs.

Much of the information Dr. Greer has gleaned has to do with either personal contacts with UFOs or personal knowledge of the government's covering up of such information. Members of the military, especially pilots, have told Dr. Greer about sightings they have had which they have not been allowed to disclose, and flight controllers have told Dr. Greer about radar contacts with unidentified objects which have been purged from the records over the years.

Part of Dr. Greer's stated objective is to confront members of congress with the evidence he has amassed so as to raise the level of awareness among individual representatives. Dr. Greer has repeatedly called for a congressional investigation into the government's dealings with UFOs, complete with subpoena power for the relevant committees as protection to prospective witnesses and to obtain necessary documents to substantiate such contacts. *See* **Greer, Stephen**

Disinformation

A deliberate strategy to provide false and misleading information to one's adversary so as to cause him to rely on that information or to send him on a wild-goose chase. Often a disinformation program involves the release of enough truthful information to allow verification, but combined with salient false facts.

According to the UFO community, there have been many disinformation programs used by the American government. The Haines CIA report admits to an active disinformation program to use UFOs as a cover for secret military weapons testing.

That same report also admits to the CIA using disinformation programs against members of the UFO community to discourage them from pursuing their research. There are still some who believe that the original MJ-12 documents were part of such a disinformation program, and that MJ-12 never really existed. *See* **Central Intelligence Agency, History of UFOs in; Debunkers; Majestic; Shandera, Jaimie**

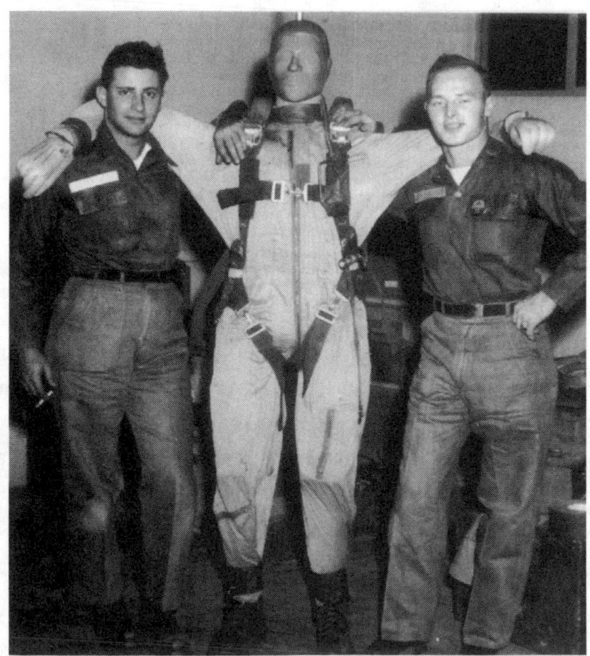

Two unnamed Air Force personnel support a beleaguered crash dummy—perhaps retrieved from the crash of a balloon near Roswell.

DNA

The year 2003 marked the fiftieth anniversary of Francis Crick and James Watson's discovery of DNA and their work in attempting to discover the processes involved in cellular development. For UFO researchers, particularly those who believe that life on earth was seeded by an alien race, DNA research holds the promise that their theories might actually be proven correct. At least one theory, while unproven as yet, involves the cloning and hybridization of the human species by an organization called Clonaid. DNA research is also central to the studies being conducted into the claims of Australian aborigines who say they have been the subject of many alien abductions, medical experimentations, and hybridization.

Some abductees have claimed there is a gradual transfer of DNA going on during the hybridization process in which various genetic strains of human beings are being reengineered for adaptation to living on other planets when earth's ecosystem fails. This, the aliens have told their abductees, is a way to ensure the survival of the human species as it colonizes other worlds. *See* **Clonaid; Darwinism; Raelians**

DoD

See **Department of Defense (DoD)**

Dolan, Richard

Author of *UFOs and the National Security State.*

Dowding, Lord

British Air Chief Marshal Lord Dowding was quoted in the July 11, 1954 *Sunday Dispatch*:

> More than 10,000 sightings have been reported, the majority of which cannot be accounted for by any scientific explanation ... I am convinced that these objects do exist and that they are not manufactured by any nation on earth. I can therefore see no alternatives to accepting the theory that they come from extraterrestrial source.

Drake Equation

A formula developed by Frank Drake in 1961 to calculate the relationships among the factors which determine how many intelligent, communicating civilizations there are in our galaxy. As Bradley Keyes explained in an article in *UFO Magazine:*

R represents the number of stars in the Milky Way galaxy that are created each year; *fp* is the fraction of stars that have planets around them; ne is the number of planets per star that are capable of sustaining life; *fl* is the fraction of planets in ne where life evolves; *fi* is the fraction of *fl* where; *fc* is the fraction of *fi* that communicates; *L* is the fraction of the planet's life during which communicating civilizations live. When all of these variables are multiplied, *N* equals the number of communicating civilizations in the galaxy.

Dreamland

Even though the government officially claims no knowledge of the goings on at the Dreamland site of Area 51 and denies the existence of such a facility, those who have tried to visit Dreamland to get a glimpse of what Bob Lazar says is a sophisticated workshop utilizing reverse-engineered ET technology are threatened by a Wackenhut security force with the authorization to shoot interlopers on sight. *See* **Area 51**

An official Area 51 insignia for a facility that is not on any map.

DSP

See **Defense Support Program**

Druffel, Ann

Author, researcher, and sociologist whose work on UFO reports from the Southern California area began in 1957 with the National Investigations Committee on Aerial Phenomena (NICAP)). After NICAP's demise in 1970, she became active in J. Allen Hynek's Center for UFO Studies (CUFOS) and the Mutual UFO Network (MUFON).

In 1980 she released *The Tujunga Canyon Contacts*, co-authored with parapsychologist D. Scott Rogo, and in 1998 wrote *How To Defend Yourself Against Alien Abductions.* In her work, she characterizes the alien abduction scenario as a manifestation of historical tales of faeries, Muslim jinn, incubi, and other multicultural expressions of deceptive, shape-shifting entities. She believes such entities may be interdimensional.

Her 2003 book *Firestorm, Dr. James E. MacDonald's Fight for UFO Science,* chronicles the involvement of a mainstream physicist and university professor in UFO research. Ann Druffel is on the Board of MUFON Los Angeles. *See* **Alien Abductions; Center for UFO Studies (CUFOS); Mutual UFO Network (MUFON); National Investigation Committee on Aerial Phenomena (NICAP)**

DuBose, Thomas J.

In the now-famous J. Bond Johnson photograph of Gen. Roger Ramey examining the Roswell crash debris, Col. Thomas DuBose is sitting to his right. In a 1991 interview with the newspaper *Florida Today,* the retired officer said that Ramey replaced the real Roswell wreckage with that of a weather balloon, and that's what he showed to the press. DuBose said that he had transferred the real Roswell crash debris, encased in a lead-lined pouch, to Washington, D.C., on orders from Gen. Clements McMullen, who also ordered him to "make up" a story to "get the press off our backs." *See* **Johnson, J. Bond; Ramey, Roger**

Dwyer, Steve

A firefighter with the Roswell Fire Department who was among the first rescue personnel to arrive at the Roswell crash site and observe alien bodies and the actual debris. Dwyer told his family that one of the aliens was carried away on a stretcher by military personnel, and that he could tell it was dying by looking into its eyes. He said that the alien telepathically communicated to him a feeling of intense sorrow. *See* **Roswell, New Mexico**

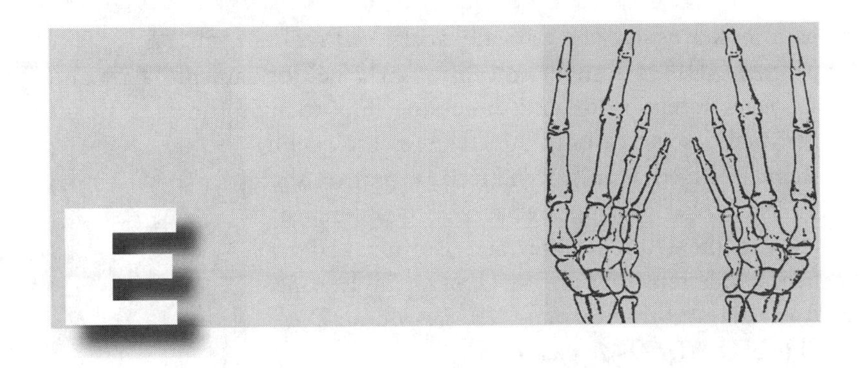

Easley, Edwin

Reportedly, the officer commanding the military police unit guarding the site of the Roswell crash. Maj. Easley ordered his men to surround the spacecraft with their backs to it and not to let anyone approach unless they had clearance. *See* **Roswell, New Mexico**

EBE

See **Extraterrestrial Biological Entity**

Echelon

Top secret National Security Agency intelligence-gathering operation with listening posts all over the world for filtering communications between millions of people, organizations, governments, and foreign military. The NSA can intercept approximately three billion communications per day.

Ecker, Don

Don Ecker has been the director of research for *UFO Magazine* since 1988 and is married to *UFO Magazine's* Editor-in-Chief Vicki Cooper Ecker. A medically retired law enforcement officer, he has applied a skeptical but honest eye on the UFO subject. In 1991 he broke the Soviet Phobos II story on CNN and in 1992 he reported on the STS-48 shuttle encounter for NBC and CNN.

During the 1990s he wrote the "Whistleblower" series for *UFO Magazine* that detailed the stories and cases of such notable UFO personalities as John Lear, Bob Lazar, Bill Cooper, and Lee Graham. Ecker hosted the critically acclaimed radio program "UFOs Tonite!" and "Strange Daze" for six years, interviewing and discussing important personalities and cases in the UFO research field. Acknowledged as an international expert on the subject, Ecker has appeared at UFO conferences both in the United States and Europe, and on nationwide television in the U.S., Japan, Europe, Australia, and the Russian CIS. *See* **Moon, Dark Side of the; *UFO Magazine***

Don Ecker, UFO Magazine's *Director of Research*

Ecker, Vicki

Before marrying researcher Don Ecker, freelance writer Vicki Cooper copublish and edited *UFO Magazine* with former Los Angeles journalism colleague Sherie Stark. Over the years the magazine went through a number of changes in staff and management, while Ecker remained editor-in-chief, a position she's maintained since 1986.

Vicki Ecker, UFO Magazine's *Editor in Chief*

Ecker continues to guide the editorial direction of *UFO Magazine* from the standpoint of fairness and accuracy, often contributing her own writings to the publication along with husband Don. Her personal take on the UFO phenomenon initially embraced a strong spiritual component, and while that remains fundamental, she believes that a gradual realization that there is no fixed answer to the mystery became the wisest attitude to take in editing the magazine.

Of all the UFO theorists who have endeavored to plumb the paradoxes of the UFO entity, Ecker most appreciates Jacques Vallee, whose viewpoint on UFOs as a method of social control best captures the underlying essence of the phenomenon. *See* ***UFO Magazine***

Edison, Thomas

The American inventor and founder of General Electric, whose designs for the light bulb, the phonograph, the direct-current (DC) power-generating system, and the motion picture camera are still in use today. Edison's direct relationship to UFO issues concerns his invention of the automatic pen. This device copied one's signature and duplicated it on documents with such precision that each signature looked exactly alike. One of the criticisms leveled by skeptics of President Truman's signature on the MJ-12 documents is that it is exactly like other signatures of Truman. This lack of variation from document to document suggests that the signature was faked and photocopied onto the document. However, if Truman were using an automatic pen, his signature would be exactly alike on all documents—thus the skeptics' argument is potentially rebutted. *See* **Tesla, Nikola**

Edwards Air Force Base, California

Reportedly the site of alien spacecraft landings, as observed by American astronaut Gordon Cooper in 1951. Edwards is also the West Coast base for Space Shuttle landings and the development site for some of America's most secret aircraft, including the B-2 Stealth bomber. Edwards Air Force Base (or rather, Muroc Field, before it was renamed) was the purported site of a meeting between President Dwight D. Eisenhower and extraterrestrials in 1954.

Eisenhower, Dwight D.

The thirty-fourth president of the United States and former Supreme Allied Commander during World War II, Eisenhower was the purported recipient of a briefing report on UFOs drafted by the group known as MJ-12, a study team formed by President Truman after the 1947 Roswell crash. Eisenhower also reportedly started Project Gleem, which later became the CIA's Project Aquarius, to study UFOs.

Perhaps the most intriguing story about President Eisenhower and UFOs concerns the alleged meeting he had with a group of aliens at Edwards Air Force Base (Muroc Field) in 1954, at which the aliens expressed the wish to disclose their presence to humanity and to bring a new age of enlightenment and scientific progress, as well as helping humanity understand its relationship to the rest of the universe. But Eisenhower told them that humans weren't ready for that kind of disclosure, that they would panic at the thought of an alien presence on earth, resulting in the collapse of social and governmental institutions.

Perhaps following their own prime directive, the aliens acceded to the president's wishes, although reluctantly, but told him they would remain a presence on earth, observing, conducting experiments, making occasional contacts with human beings, and establishing their own facilities so as to educate humanity for the day when they would reveal their own presence. Reportedly, Eisenhower agreed to cooperate on that basis, and a deal was struck. The aliens would select groups of human beings to prepare for disclosure, while the world's governments would study aliens and their technology. *See* **Majestic; Projects, Government UFO**

El Baz, Farouk

A lunar geologist who offered the theory that there are huge caverns beneath the moon's surface. Although experiments conducted during the Apollo missions suggested just such a "hollow moon," results of these experiments have never been publicly disclosed. *See* **Moon, Dark Side of the**

Electromagnetic Envelope

An envelope or "cushion" of energy that is said to surround flying saucers, allowing them to navigate through space. By manipulating the shape of the electromagnetic envelope, alien pilots are able to make their craft perform maneuvers that would tear apart conventional aircraft and accelerate at speeds that would create near fatal G-forces for human beings.

Electromagnetic Force Field

Supposedly a protective envelope UFOs use to prevent intruders from approaching, or a field automatically generated by the craft itself as part of its propulsive process that seems to create what UFO researchers call the *Oz effect,* perceived as a slowing down of time. *See* **Oz Effect**

Energy Weapons

Directed-energy or particle-beam weapons that shoot high-speed streams of electrons at their targets. Such weapons, possibly found in the wreckage of the Roswell craft, destroy electrical circuitry, disable navigation and computer-control systems, heat the surface of incoming warheads (thereby leading to their detonation), and even raise the temperature of human skin to intolerable levels so as to disable groups of enemy troops. Energy weapons were part of the Star Wars Defense program first proposed by President Reagan. *See* **Strategic Defense Initiative (SDI)**

ET

See **Extraterrestrial**

"ET Exposure Law"

Before the July 1969 Apollo moon landing, NASA was given, presumably by executive order, the right to quarantine under armed guard anyone coming into contact with ETs. The publicized reason for this was to protect against contamination from alien microbes.

Exeter, New Hampshire

The 1965 UFO police sighting near the town of Exeter in New Hampshire in which local law enforcement officer Eugene Bertrand, after first having been apprised by a female driver that she had been buzzed by a circular object itself encircled by a bright red light, later learned that a local hitchhiker, making his way home to Exeter also saw a similar object enveloped in a red light.

The hitchhiker, an eighteen-year-old named Norman Muscarello, observed the object hover over a nearby farmhouse and after trying to rouse the occupants of the house, made his way back to Exeter where he contacted local police who then alerted units about the young man's sighting. The officer who had encountered the female driver who'd first witnessed the object returned to the police station to retrieve Muscarello so that he could show police the location of the sighting.

At the location, Bertrand and Muscarello exited the police car and walked into the field towards the farmhouse where Muscarello saw the object ascend from behind a treeline. The object remained in sight while two men returned to the police car where Bertrand radioed in his report.

Another unit arrived on the scene, driven by David Hunt, and he, too, observed the object with Bertrand and Muscarello as it moved back and forth in the sky and executed hairpin turns and maneuvers. They were able to observe individual lights on the side of the object. The object soon disappeared, but at least one additional report concerning its appearance had been made by a witness from a pay phone to an operator in Hampton, New Hampshire.

The incident was investigated by personnel from the local Pease Air Force Base near Portsmouth, New Hampshire, who tested the area for radioactivity, but found none. The incident drew news coverage from local papers as well as the *Manchester Union* and was the subject of a book by Raymond Fowler and John Fuller called *Incident at Exeter.*

Exobiology

The study of biology as it pertains to possible extraterrestrial life forms. Although the U.S. government has officially denied the existence of UFOs, it has provided NASA the necessary budget funds for exobiology studies, including monies for field trips to those parts of our planet where conditions are so extreme that they mimic conditions on other worlds.

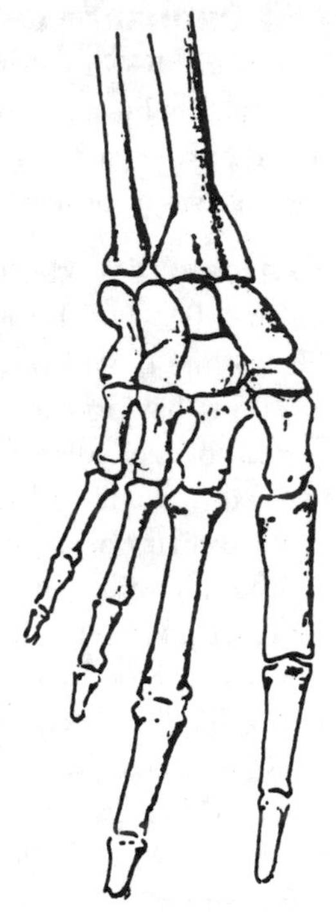

An artist's conception of the bone structure of an alien hand.

"Experiencers"

Those individuals who have been abducted by, or have had encounters with, or have been contacted by or have communicated with by extraterrestrials. The name can also apply to long-term contactees or members of families who possibly have undergone multigenerational testing or hybridization. It is also not uncommon for therapists to refer to their clients as experiencers, rather than patients, so as to remove the stigma of being violated. *See* **Abductees; Contactees**

Extraterrestrial Biological Entity (EBE)

The term first appeared in the MJ-12 documents, and refers to the alien creatures recovered at Roswell.

Extraterrestrial (ET)

The designation first popularized in Steven Spielberg's film *ET, The Extra-Terrestrial*, referring to life forms from planets other than earth—although that life need not be "intelligent" or "sentient" as we understand it.

Many scientists who are convinced that extraterrestrial life exists are not necessarily adherents to the idea of flying saucers or the aliens usually associated with them. These scientists tend to believe that extraterrestrial life can take any form, from primitive single-celled entities to strains of bacteria that can survive under the most hostile of conditions, to actual sentient beings.

Other names for extraterrestrials are *EBEs* and *aliens,* which have been used interchangeably. *See* **Alien Beings; Grays; Traders; Zetas**

Ezekiel

In one of the most dramatically vivid sections of the Old Testament, the prophet Ezekiel describes his encounter with beings who seem to descend from a "wheel of fire" turning in the sky. The "living creatures" had the appearance of "burning coals of fire, and the appearance of torches," who "ran and returned as the appearance of a flash of lightning." UFO researchers who believe the Old Testament is a retelling of the story of how sentient life was brought to earth by extraterrestrials seize upon the Book of Ezekiel as evidence of UFO contact. The "wheel of fire in the sky turning around upon itself" is a flying saucer, and the living creatures who confront Ezekiel with prophecies of the future and directions to correct the transgressions of the children of Israel are extraterrestrials. *See* **Däniken, Erich von**

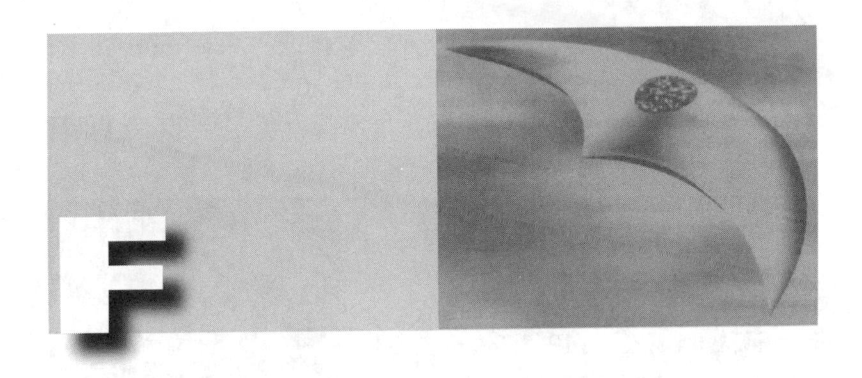

FAA

See **Federal Aviation Administration**

Face on Mars

See **Mars, Face on**

False Memory Syndrome (FMS)

The phenomenon whereby someone in authority can, inadvertently or deliberately, implant false memories into the mind of a subject. As demonstrated by the experiments conducted by Professor Elizabeth Loftus, who was attempting to prove that even eyewitness testimony sometimes cannot be trusted, subjects were interviewed by questioners who deliberately suggested events that did not and could not possibly have taken place. The subjects were convinced, afterwards, that they remembered such events.

False memory syndrome is a controversial subject in UFO research because some skeptics believe it is the regression therapists themselves who are preconditioned to believe that events reported by abductees are, in fact, alien abductions, and who then suggest that scenario to their vulnerable and highly suggestible patients. *See* **Abductees; Abduction Scenarios**

Fastwalkers

The codename for UFOs picked up by U.S. surveillance satellites. Sources who work inside the defense industry say such satellites are detecting fastwalkers at a rate of three to four every month. The

Defense Department describes them as a "source of electromagnetic radiation moving at high speed in the outer layers of the atmosphere, which triggers the sensors of spy satellites." *See* **Defense Support Program (DSP)**

Fate Magazine

One of America's longest-running publications reporting on UFOs, extraterrestrials, ghosts, spirits, and other paranormal issues. In 1962, in an incredible interview for that time, Dr. Hermann Oberth, a founder of the modern age of rocketry who came to the United States after World War II to work on America's guided missile programs, told *Fate* that not only do UFOs exist, but they propel themselves and navigate by generating "artificial fields of gravity." *See* **Oberth, Hermann**

An early Fate *magazine cover.*

FBI

See **Federal Bureau of Investigation**

Federal Aviation Administration (FAA)

The government agency primarily responsible for the regulation of civil aviation, establishing airline safety procedures, and pilot and flight-crew standards and licensing. The majority of commercial pilots who have had UFO sightings routinely refuse to file reports because the FAA frowns on such disclosures, and a pilot's official admission that he has seen a UFO can ruin a career.

Federal Bureau of Investigation (FBI)

Although the federal government has officially denied the existence of UFOs and any involvement in the subject, FBI Director J. Edgar Hoover insisted in a series of memos, uncovered during Freedom of Information Act requests by UFO researchers, that his agents find out what really happened at Roswell.

In a 1947 note to his assistant Clyde Tolson, regarding a request that the FBI take over the investigation of possible UFOs, Hoover wrote, "I would do it, but before agreeing to it we must insist upon full access to discs recovered. For instance in the [second] case the

Army grabbed it and would not let us have it for cursory examination." Hoover was reacting to memos from his field supervisors who reported that the Air Force had retrieved crashed flying saucers in New Mexico.

Federal Emergency Management Agency (FEMA)

An independent agency reporting directly to the president and tasked with responding to, planning for, recovering from, and mitigating against disaster. FEMA was created by President Jimmy Carter's 1979 executive order merging many of the separate disaster-related responsibilities into a self-contained agency. There are wide fears that FEMA has the authority to suspend Constitutional guarantees of civil rights after a disaster is declared. It is also believed that FEMA is part of the "black" budget, and can launder funds into covert operations.

FEMA's role in what some have called a "secret government" came to light only after Hurricane Andrew smashed into Florida in the early 1990s. A Congressional review revealed that the agency had spent more money for so-called "black operations" than for Florida disaster relief. So secret are some of FEMA's facilities and operations that one would stand a better chance of penetrating a nuclear missile silo than FEMA's Mount Weather facility in Virginia.

Under Department of Defense Directive 5525.5, entitled "DoD Cooperation with Civilian Law Enforcement Officials," FEMA also has the ability to borrow, have transferred to it, or otherwise co-opt practically any Department of Defense asset.

In October of 1984, as President Reagan was in the midst of his reelection campaign, journalist Jack Anderson discovered that FEMA officials had drafted "standby legislation" to present to Congress if the United States was faced with domestic chaos or a state of total war. The proposal, according to Anderson, would have stripped away the essentials of U.S. democracy; it would "suspend the Constitution and the Bill of Rights, effectively eliminate private property, abolish free enterprise, and generally clamp Americans in a totalitarian vise."

It was also reported that Lt. Col. Oliver North, the Reagan White House aide who stood at the center of the Iran-Contra scandal, had worked with FEMA on top-secret projects such as military exercises designed to test the government's capacity to round up refugees and rabble-rousers.

FEMA

See **Federal Emergency Management Agency**

Ferret Satellites

Part of the U.S. surveillance satellite system, "ferrets" are orbiting electronic eavesdroppers tasked with collecting signals intelligence (SIGINT) from radar, radio, short-wave, and cell-phone communications. They have also routinely picked up signals from "fastwalkers" traveling through earth's atmosphere. *See* **Fastwalkers**

Filer, George

A former U.S. Air Force pilot who writes a weekly Internet report called *Filer's Files.* The Eastern regional director of MUFON, he has investigated sightings for over twenty years. His personal experience with a UFO took place over England during his service as an Air Force pilot when he chased the object across the sky.

Fire Officer's Guide to Disaster Control

A privately published handbook for handling the emergency procedures concerning the crash of a UFO and/or the retrieval of extraterrestrial biological entities (EBEs).

Firmage, Joe

A millionaire dotcom entrepreneur who stepped down from his company to pursue UFO studies. Firmage formed a private foundation called the International Space Sciences Organization. Some in the UFO community believe that Firmage's plan is to raise money for funding private research into technologies that have come from extraterrestrials ushering in the next wave of high-tech development expected to sweep the United States.

Joe Firmage

Fish, Marjorie

An astronomer who took Betty Hill's star map and converted it into a three-dimensional model to see if she could determine where the Hills' alien abductors had come from. Betty Hill had drawn the map under hypnotic regression, based on what she had seen and heard while onboard the alien spacecraft. When a new catalog of stars, the *Gliese Catalog,* was released, it allowed Fish to assign accurate distances between stars, and she was able to come up with a version of Hills' map that seemed to make sense. The map indicated that the beings had originated from a star called Zeta Reticuli located relatively close to our solar system. *See* ***Gliese Catalog;*** **Hill, Barney and Betty**

509th Bomber Squadron

The Army Air Force squadron based at the Roswell Army Air Field and commanded by Col. William Blanchard. The 509th, according to its own press release, recovered a "flying disc" that "landed on a ranch near Roswell." The 509th's intelligence officer, Maj. Jesse Marcel, accompanied rancher Mac Brazel to the Foster Ranch debris field to survey the area and report back to his boss, Col. Blanchard. The 509th was also, at the time, the only operational nuclear bomber squadron on earth, and had been responsible for dropping the first atomic bombs on Japan during World War II. *See* **Roswell, New Mexico**

Flight 9, Project Mogul

The latest cover story explaining the object that crashed in Roswell in 1947, replacing the original weather balloon story. In 1994, partly in response to inquiries by the late Representative Steven Schiff, the Air Force announced they had gone back and checked their Roswell files and concluded that the debris Mac Brazel found was a top-secret reconnaissance balloon, part of a project called Mogul.

Moguls consisted of a multiple-balloon array tethered together. Some of the arrays were quite long, stretching over five hundred feet. Their mission had been to monitor any attempts the Soviet Union might make to detonate a nuclear device. Records showed that a number of Moguls had been launched during the Roswell timeframe.

However, the Air Force's own report indicated that many of the balloon missions had, in fact, been scrubbed due to inclement weather. Virtually all the balloons that were successfully launched at that time were reported as recovered or located. Only one flight, Flight 9, remained unaccounted for. In addition, records of that flight were missing.

Roswell skeptics believe it was Mogul balloon Flight 9 that Mac Brazel found, but many UFO researchers disagree. First of all, the amount of debris reportedly found at the Foster Ranch exceeded any amount that would be left by a Mogul balloon array. Secondly, when one looks at the behavior of the military in the days and weeks following the crash, the extraordinary threats made against civilians went far beyond those used against any other group involved with other classified projects in U.S. history, including the development of the atomic bomb. This would have been ridiculous "overkill" for any cover story, including one involving a flying saucer—unless the saucer was real.

The crashed Project Mogul device was probably similar to this Air Force Vee balloon photographed at Holloman Air Force Base, c. 1965.

Also, the testimony of dozens of high-ranking military officers who were familiar with weather balloons, including Mogul balloons, have stated unequivocally that the debris they handled was not from any balloon; rather, it was unlike any other material they had seen before. Therefore, the Mogul story simply does not hold up under the weight of the evidence. See **Projects, Government UFO; Roswell, New Mexico**

FMS

See **False Memory Syndrome**

Flying Saucers

A term coined by the media, after Kenneth Arnold described the crescent-shaped objects he saw flying over the Cascade Mountains as "skipping" through the air much like a saucer would skip over water. Although today the objects are mostly referred to as UFOs, newspapers from the late 1940s through the 1960s consistently referred to these disk-shaped craft as flying saucers, and they have become an almost metaphorical image representing the strange, the exotic, and the unknown.

Dr. Carl Jung referred to these objects as flying saucers and said the circular shape was an archetype embedded in humanity's collective unconscious. *See* **Arnold, Kenneth**

Flying Shields

During the Dark Ages, after the fall of Rome when Germanic tribes were spreading across Europe, various tribal chronicles reported sightings of flying ships and "flying shields." Perhaps the most colorful of these accounts came from *Annales Laurisseness,* about an event during a siege of a castle in 900 CE, in which the chronicler writes:

> Those watching outside in that place, of whom many still live to this very day, say they beheld the likeness of two large shields, reddish in color, in motion above the church, and when the pagans (probably pre-Christian Vikings raiding the Saxon enclaves) in who outside saw this sign, they were at once thrown into confusion and, terrified with great fear, they began to flee from the castle.

Flying Triangles

One of the more mysterious phenomena in ufology, flying triangles have been observed by thousands of witnesses, over heavily populated metropolitan areas as well as isolated desert communities—all this to the absolute denial of their existence by the U.S. military and senior political leaders.

Flying-wing or flying-triangle designs for aircraft have been around since the 1920s, when Alfred Lippisch designed a wing-shaped glider in Germany. During the 1930s, the Germans experimented with flying-wing designs, trying to develop jet-propelled bombers. Flying wings also appeared in *Things to Come,* the 1936 movie based on H.G. Wells' *The Shape of Things to Come.* The Horten brothers created a jet-propelled flying delta for the Luftwaffe in 1945, which never saw action.

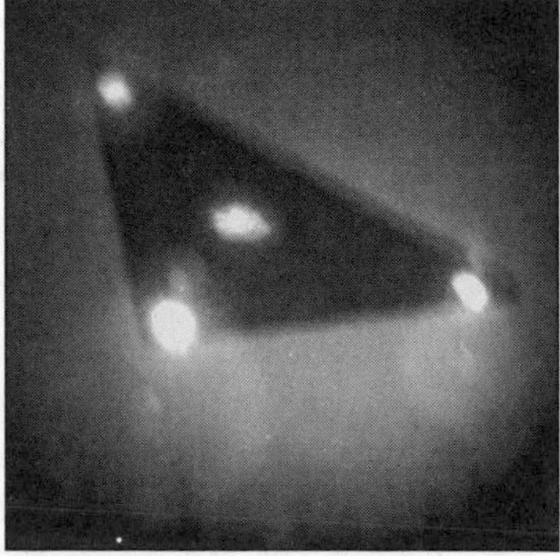

These craft have been seen over New York's Hudson Valley and in the skies over Phoenix, Arizona.

In the early 1950s there were sightings of flying triangles, most likely designs based on Northrop's YB-49. Flying wings remained an experimental military design, finally becoming operational with the B-2 Stealth Bomber at the end of the 1980s.

During the 1980s, flying triangles were observed cruising up and down New York's lower Hudson Valley, even over Manhattan itself. The huge, 747-sized craft made no noise as they floated by. The sightings have continued unabated for over twenty years, including headline-making sightings of flying triangular objects over Phoenix, Arizona in 1997. They have also been spotted over European countries, particularly over Belgium in 1989-90, where jet fighters sought to intercept them, but to no avail.

Although most government and military insiders are closed-mouthed about flying triangles, at least one aerospace executive has reported that triangular-shaped, neutral-buoyancy super blimps have been in the works at Lockheed for years. These craft are truly spectacular in size, having a wing/fuselage area of over a thousand square feet. Their undercarriages are painted with a nighttime camouflage so that ground observers might think they're looking at fuzzy stars instead of objects. The blimps can remain aloft for days, hover, accelerate, and move up and down, all while remaining absolutely silent—thanks to their electrically polarized aerodynamics and navigation system, which also emits a static electricity field that can create the *Oz effect,* causing people to feel disoriented.

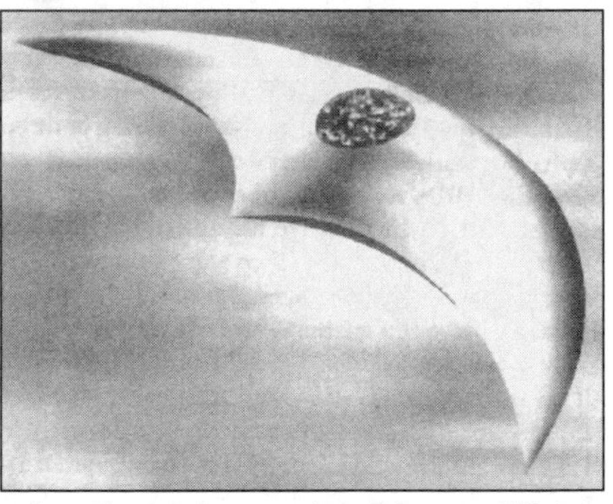

An early artist's conception of the sort of "flying crescent" that was described by Kenneth Arnold in the 1940s.

This super blimp, some aviation reports have said, was part of a development program aimed at transporting large numbers of troops from their home bases to the front lines in less than twenty-four hours. The blimp was also designed as a high-tech offensive weapons platform for launching air-to-surface missiles or directed-energy weapons while at the same time protecting itself from attack. It can fly at altitudes so high as to be almost beyond the range of any surface-to-air missile. In its peace-time incarnation, the blimp could function as the ultimate passenger airliner or a commercial transport for companies like Federal Express or UPS. *See* **Phoenix Lights; Oz Effect**

FOIA

See Freedom of Information Act

Foo Fighters

A rare photo of Foo Fighters in action over Germany in the 1940s.

Strange, brilliantly colored balls of light that World War II bomber pilots commonly observed maneuvering around their formations during raids over Germany. Allied pilots believed the balls of light were secret German weapons. However, Luftwaffe squadrons also saw the objects and feared they were a secret Allied tracking device.

Ford, Gerald R.

The thirty-ninth president of the United States. In 1966, while still a Michigan congressman, Ford wrote to Mendel Rivers, chairman of the House Armed Services Committee, saying, "In the firm belief that the American public deserves a better explanation than thus given by the Air Force, I strongly recommend that there be a committee investigation of the UFO phenomenon. I think we owe it to the people to establish credibility regarding UFOs and to produce the greatest possible enlightenment on the subject."

Foreign Technology Desk

In the Army Office of Research and Development at the Pentagon, this division was responsible for the investigation of technology originating from sources other than those in the United States, for the purposes of acquiring that technology for the U.S. Army. This is the desk to which Lt. Col. Philip J. Corso was assigned in 1961 by Lt. Gen. Arthur Trudeau, and from which he began his investigation into the material in the Army's Roswell File, consisting of a laser cutting tool, an atomically aligned piece of fabric, micro solid-state circuitry, a fiber-optic junction, a night-vision light-collecting inner eyepiece, and a special headband transceiver used to pick up electronic waves from the brains of the ET pilots of the Roswell craft. Corso revealed in *The Day After Roswell* that he brought these technologies to American industries for reverse-engineering. *See* **Corso, Philip J.**

Forrestal, James

Undersecretary of the Navy, then Secretary of the Navy during World War II in the Roosevelt administration, Forrestal became America's first Secretary of Defense in 1947 in the Truman administration. Forrestal was also one of the alleged members of MJ-12, according to researcher Stanton Friedman. In 1949 Forrestal was admitted to Bethesda Naval Hospital after suffering a nervous breakdown. Shortly thereafter, he committed suicide. Members of the UFO community have speculated that Forrestal's breakdown was directly related to the disturbing information he was privy to as a member of MJ-12, and others have even suggested that he was murdered to protect the Roswell cover-up. *See* **Majestic**

James Forrestal

Fort Riley, Kansas

A military base where, reportedly, the bodies of extraterrestrials were stored in 1964. Fort Riley was also the site where in 1947 a cache of material from the Roswell crash was stored on its way to Wright-Patterson. It was at Fort Riley in July 1947 that Philip Corso, then a young officer in intelligence school and that night's duty officer, learned that some strange cargo had just arrived from New Mexico. Looking inside one of the shipping crates from Roswell, Corso saw the body of an extraterrestrial. It wasn't until he reached the Pentagon in 1961 and was handed the Roswell file that Corso read the autopsy report from Walter Reed Army Hospital and realized that the autopsy conducted on the Roswell aliens might very well have included the creature that Corso saw that night at Fort Riley. *See* **Corso, Philip J.**

Foster Ranch

Site of the debris field in New Mexico where ranch foreman Mac Brazel discovered the crashed remains of a UFO which are later retrieved by an Army team and taken back to the Roswell Army Air Field. *See* **Brazel, William "Mac"; Roswell, New Mexico**

Fowler, Raymond

A UFO investigator and author who has written about the abduction cases of Betty Andreasson *(The Andreasson Affair, The Andreasson Affair, Phase Two)* and Jim Weiner *(The Allagash Affair)*, as well as *Incident at Exeter.*

Prolific author and UFO researcher Raymond Fowler.

Fowler served as chairman of the NICAP Massachusetts subcommittee, an early warning coordinator for the USAF-contracted study at the University of Colorado, and as a scientific associate for the Center for UFO Study. In later years he served as director of investigations on the board of directors of the Mutual UFO Network. Dr. J. Allen Hynek has called Fowler, "an outstanding UFO investigator. I know of none who is more dedicated, trustworthy or persevering."

Freedom of Information Act (FOIA)

One of the most powerful weapons in the arsenal of UFO researchers. Used to dig information out of government records, the 1974 act was signed into law by President Gerald Ford in the wake of Watergate and the Pentagon Papers scandals.

Three researchers in particular whose work has been the result of persistent and oftentimes very clever FOIA searches are Richard Dolan, whose book *UFOs and the National Security State* is one of the more important books in modern UFO scholarship; John Greenewald Jr., whose "Black Vault" website is a virtual treasure-trove of UFO-related documents; and Clifford Stone. Retired Army Specialist Stone claims he was a member of a UFO retrieval team in the military, sworn to secrecy under the National Security Act, but whose subsequent FOIA searches for documents relating to the missions to which he was purportedly assigned revealed the existence of Operations Moondust and Bluefly, both of which involved the retrieval of downed space objects and the cataloging of information about these objects. *See* **Dolan, Richard; Greenewald Jr., John; Stone, Clifford**

Freeman, George P.

The Air Force spokesperson for Project Blue Book at the Pentagon who expressed concern about reports of "Men in Black" who were silencing UFO witnesses. Col. Freeman wrote:

> Mysterious men dressed in Air Force uniforms or all in black and bearing impressive credentials from government agencies have been silencing UFO witnesses. We have checked a number of these cases, and these men are not connected to the Air Force in any way. We haven't been able to find out anything about these men. By posing as Air Force officers and government agents, they are committing a Federal offense. We would sure like to catch one—unfortunately the trail is always too cold by the time we hear about these cases, but we are still trying.

French Flying Saucers Report (COMETA)

In 1998 a remarkable report was released by former French defense officials, which for the first time revealed evidence that top officials in the French military had direct knowledge of the existence of UFOs and which also suggested that UFOs piloted by extraterrestrials might be hostile.

Written by a private association called COMETA (Committee for In Depth Studies), the document was titled "UFOs and Defense. What must we be prepared for?" ("Les OVNI et la Défense. A quoi doit-on se preparer?") and was the result of an in-depth study of UFOs covering many aspects of the subject, especially questions of defense.

The report looked at the possibility that the United States or other countries may have entered into deals with the aliens who pilot these UFOs, and speculated what that might mean for a future conflict with these beings.

The significance of the report is not just that the members of the committee evaluating the evidence had outstanding credentials, it was that the report made it all the way to the very top of the French military, where those at the highest levels of command signed off on it as an accurate representation of the state of affairs.

Friedman, Stanton T.

Stanton Friedman

Nuclear physicist Stanton Friedman has been investigating and documenting the existence of flying saucers and extraterrestrial beings for over thirty years. He has written several books on the subject and has lectured at over a thousand colleges and universities. Friedman was the person who initially uncovered the Roswell crash story. As a trained scientist, he has brought a deliberate and thorough approach to researching this complex subject.

In 1978, while Friedman was being interviewed at a radio station in Baton Rouge, Louisiana, the director of the radio station suggested that he ought to talk to Jesse Marcel, the military man who had handled pieces of the crashed saucer. Friedman contacted Marcel, and that was how the Roswell story finally went public. In addition to his work on the crash at Roswell, Friedman's book on MJ-12 (*Top Secret/Majic*) and the individuals who reportedly served on that group has become the definitive study of the first organized government cover-up of UFOs and the origins of that cover-up. *See* **Majestic**

Fuller, John G.

Co-author, with Raymond Fowler, of *Incident at Exeter. See* **Fowler, Raymond**

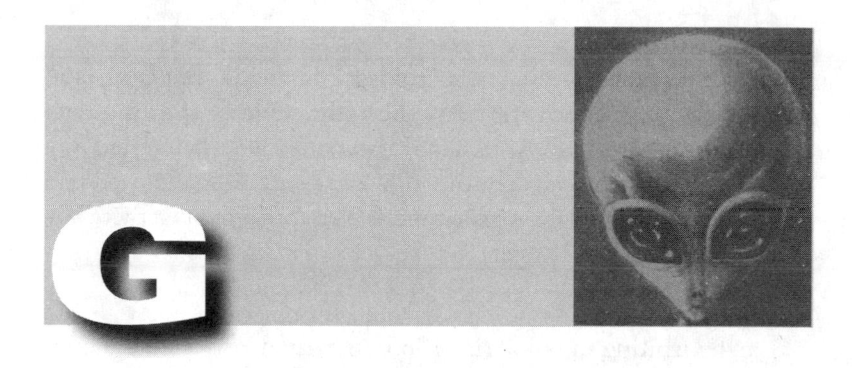

Gairy, Eric

Prime Minister of Granada, who proposed to the UN that the year 1978 be declared "United Nations Year of the UFO." He also urged the UN to establish an agency for psychic research into UFOs and the Bermuda Triangle. One year later, his government was overthrown by his political rivals, who used his interest in UFOs to disparage his mental state.

Galley, M. Robert

The former French Minister of Defense, who, in a 1974 radio interview about reports of UFO sightings by military and civilian pilots, told the listening audience:

> I must say that if listeners could see for themselves the mass of reports coming in from the airborne gendarmerie charged with the job of conducting investigations, all of which reports are forwarded by us to the National Centre for Space Studies, then they would see that it is all pretty disturbing.

Gallup Poll

A survey conducted by this national polling organization in 1996 revealed that 71 percent of the American public believed that the U.S. government knew more about UFOs than they had publicly revealed. Subsequent polls have shown that more than half the American public believes that UFOs have visited earth and contacted world governments.

Gauss Meter

Device used to measure electromagnetic fields. Implants, still inside abductees who claim to have been the subjects of medical experiments on board UFOs, usually register off-scale when tested by a Gauss meter. However, when the implant is surgically removed and then tested, it no longer registers high electromagnetic activity. *See* **Implants, Alien; Leir, Roger**

Gemini 4

Earth orbiting mission in 1965 in which astronauts James McDivitt and Ed White observed a strange cylindrical object with a "protuberance" traveling in orbit with their space capsule. The astronauts also observed a bright light traveling in higher orbit. The objects were never identified. The astronauts' sighting was considered important enough to include in the study undertaken by the Condon Committee in 1969. *See* **Condon Report; McDivitt, James; White, Ed**

Gemini 7

Astronaut Frank Borman saw what he called a "bogey" flying in formation with his spacecraft. *See* **Apollo Program; Borman, Frank; Lovell, James**

GEOS Satellites

U.S. weather satellites which are part of the surveillance array that has spotted "fastwalkers" entering and maneuvering in the earth's atmosphere. *See* **Fastwalkers**

Gersten, Peter

An attorney and former prosecutor in the Manhattan District Attorney's office, Gersten investigated the Hudson River Valley flying triangle sightings of the 1980s, and founded the UFO organization CAUS (Citizens Against UFO Secrecy) in Arizona. Gersten filed a number of lawsuits against the federal government for failure to disclose information it has on UFOs and most recently after the Phoenix Lights incident in 1997, charged the government with failing to protect Arizona residents against a potentially

Peter Gersten

hostile intrusion by extraterrestrial craft. *See* **Citizens Against UFO Secrecy (CAUS)**

Ghez, Andrea

An astronomer who wrote a 1998 article in the *Los Angeles Times* in which she said that while using the Keck Telescope in Hawaii, she was able to determine that twin star systems—the type of system that Betty Hill described under hypnotic regression after her abduction—are newly-born stars and that almost all newly born stars are twin stars. *See* **Hill, Barney and Betty; Fish, Marjorie**

Ghirlandaio

The fifteenth-century artist who painted "The Madonna and St. Giovannino," with the Virgin Mary in the foreground and a flying disk emanating sparks in the background.

Ghost Rockets

From 1932 to 1937 there were hundreds of reports from Finland, Norway, and Sweden of unknown aircraft flying overhead. This was at a time just prior to World War II when aircraft were far less prevalent in the skies than they are today, especially in Scandinavia. Huge craft were reported, some of which had multiple engines, accompanied by other craft with strange multicolored lights. These "airships" were seen hovering over military installations, much as flying saucers do today. Then, just as suddenly as the sightings first appeared, they ceased.

In 1946 strange vehicles appeared in the skies over Scandinavia again. The objects' behavior did not match the flight patterns of any known rockets or aircraft. Many of the UFOs were reported to be traveling at extraordinary velocities and making hairpin turns at high speeds, or diving into lakes. Allied armed forces insisted they were not from any of their countries, leaving the question of the origin and identity of these "ghost rockets" unresolved.

Glaser, Hans

A sixteenth-century artist whose woodcut depicts a battle between flying disks and cylinders, reportedly fought on August 4, 1561 over Nuremberg, Germany.

Gleason, Jackie

The famous television and movie comedian was a close friend of then-President Richard Nixon and personally interested in UFOs.

Gleason once asked Nixon to tell him what he knew about UFOs, so the president arranged for Gleason to be flown by presidential helicopter to nearby Homestead Air Force Base, where, under heavy security, he was allowed to view the preserved bodies of several small alien beings. Gleason was apparently very shaken by the whole experience. His wife at the time, Beverly McKittrick, recalls how Gleason returned home visibly upset that night in 1973. Gleason later built a home in New York State's Catskill Mountains shaped like a flying saucer. *See* **McKittrick, Beverly**

Gleem

See **Projects, Government UFO**

Gliese Catalog

The catalog of stars that allowed astronomer Marjorie Fish to locate the twin-star system where Betty Hill said her alien abductors originated. *See* **Fish, Marjorie; Hill, Barney and Betty**

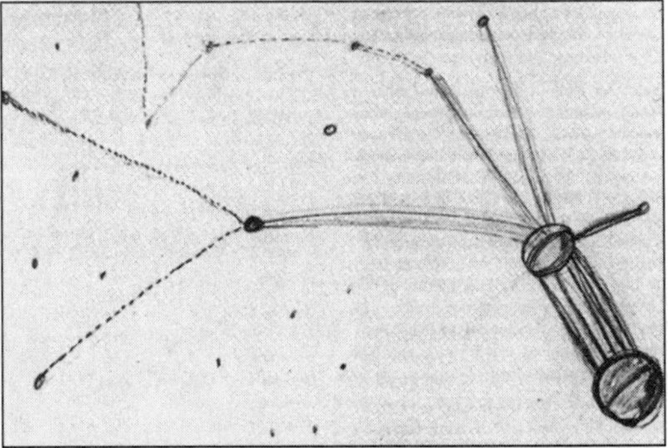

Zeta Recticuli system. Betty Hill's drawing of the home star system of the Zetas.

Godfrey, Arthur

American broadcasting pioneer who had a close-up UFO encounter in 1965, which he reported on his television show. Godfrey said that a UFO took up a position alongside the wing of his private plane, coming so close that Godfrey, in the pilot's seat, had to veer off to avoid what he thought was going to be a collision. The UFO then paced the plane and seemed to be shadowing it, maneuver for maneuver. Godfrey said the UFO encounter worried him greatly with its portent.

Godman Air Force Base, Kentucky

The air force base from which Air National Guard captain and P-51 fighter pilot Thomas F. Mantell flew on January 9, 1948 to pursue a UFO. Mantell was killed when his aircraft lost control and crashed under what some say are mysterious circumstances. *See* **Mantell, Thomas F.**

Goldin, Daniel

Daniel Goldin

NASA's former chief administrator, who was tasked with streamlining the space agency and making its missions more cost-efficient. He was heavily criticized for compromising the quality control of systems in NASA's space probes, particularly in the missions to Mars, several of which mysteriously failed.

Goldin, as head of NASA, was also considered by some UFO researchers as one of the chief custodians of the government's continuing cover-up of the UFO reality. However, after NASA reported finding what it considered a primitive form of life in a recovered Martian rock, Goldin was quoted as saying, "After that ... I excused myself ... went into my office and shut the door, and I sat there, blinds closed for half an hour, contemplating the impact that this could have on who we think we are." *See* **National Aeronautics and Space Administration (NASA)**

Goldwater, Barry

Former United States Senator from Arizona and presidential candidate, Goldwater reportedly once asked Air Force Gen. Curtis LeMay, head of the Strategic Air Command, if the Air Force had secreted away the recovered UFO wreckage and alien bodies and kept them locked away at Wright-Patterson Air Force Base's "blue room."

Not only did LeMay not answer, he sternly warned Goldwater never to ask that question again. In 1975 Goldwater wrote to a friend about his request to Gen. LeMay, saying:

> The subject of UFOs is one that has interested me for some long time. About 10 or 12 years ago I made an effort to find out what was in the building at Wright-Patterson Air Force Base where the information is stored that has been collected by the Air Force, and I was understandably denied this request.
>
> It is still classified above Top Secret. I have, however, heard that there is a plan under way to release some, if not all, of this material in the near future. I'm just as anxious to see this material as you are, and I hope we will not have to wait much longer.

Gonzalez, John N.

A ham radio operator in New Jersey who picked up the following conversation between the air traffic control tower at Newark Airport in New Jersey and Flight 262 on November 17, 1997, sometime before midnight. Crew members of two other aircraft in the area witnessed this UFO sighting and joined in the radio conversation. According to the verbatim transcript:

JET #2: Watch out! The two [UFOs] are coming up to you.

FLIGHT 262: Well, Captain, the two up here are coming down to meet with you.

TOWER: Flight 262, what is your status?

FLIGHT 262: We have 236 souls onboard and 50,000 [pounds] of fuel. I think THESE DAMN THINGS ARE GOING TO HIT US. We are over Morristown just in case there is a collision with them. (Pause) They have taken off towards the northeast. And by the way, towards the northeast, it also looks like a meteor or space debris is coming down.

TOWER: Do you wish to report a UFO sighting?

FLIGHT 262: (pause) No, we have nothing to report.

JET #2: We heard you. I am making sure the passengers are all right. And, no, I have nothing to report, either.

JET #3: You guys have seen more than your share of UFOs. I know I have.

TOWER: Who are you? Please identify yourself.

(There was no response.)

TOWER: Flight 262, go to the emergency frequency. We will meet the both of you there.

See **Newark Airport**

Gonzalez, Thomas

In 1947 at the Roswell crash site, Sgt. Gonzalez was assigned to guard the area against intruders. Years later, he recalled seeing small bodies with eyes not much larger than human eyes and a slightly enlarged head. *See* **Roswell, New Mexico**

Good, Timothy

Author of *Above Top Secret.* Good revealed that a friend of his in British military intelligence once had a conversation with a very prominent astronaut who told him that NASA had been warned by ETs to stay off the moon. *See* **National Aeronautics and Space Administration (NASA)**

Gorbachev, Mikhail

Former secretary general of the Soviet Union. In an article published in *Soviet Life Supplement* in May 1987, Gorbachev said he agreed with former U.S. president Ronald Reagan that in the event of an invasion from outer space, the United States and Soviet Union would join forces to repel it. *See* **Reagan, Ronald**

Graham, Lee

An aerospace worker who tried to authenticate the MJ-12 documents. Graham realized that anything stamped *Top Secret,* which was the stamp on the MJ-12 documents, fell under Department of Defense regulation 5200.1-R, which requires any government employee or government contract worker to report any leak of Top Secret documents. The forging or faking of any documents stamped Top Secret is also a violation of the law. Accordingly, Graham reported the MJ-12 documents to the FBI as either being stolen or forged and waited for a response.

The Bureau spent over a year investigating Graham's report to determine whether such a crime had been committed. It concluded that if the MJ-12 documents had been forged, it was the best forgery they had ever seen. The FBI also admitted that if the documents were real, they could not get any other department or agency of the government to confirm or acknowledge that the documents had been lost or stolen. Graham, by default, had forced the government to admit the documents were real. *See* **Majestic**

Grand Canyon

Site of a June 30, 1947 UFO encounter in which a Navy pilot reported having seen two flying disks land about twenty miles south of the canyon. The pilot reported that the disks had "inconceivable speed."

Gray, Gordon

Secretary of the Army during the Truman and Eisenhower administrations. According to Stanton Friedman, Gray was a member of MJ-12. *See* **Majestic**

Grays

Artist Kesara's depiction of a typical Gray alien.

The Grays (also called Zetas, since it's believed they come from the twin-star system Zeta Reticuli) are the most common aliens reported by abductees. Grays are described as short, thin-limbed and slight-bodied, with large heads and huge, almond-shaped black eyes, and gray skin color. *See* **Alien Beings; Zetas**

Great Airship Mystery

During the period of late 1896 through early 1897, various types of self-propelled, apparently lighter-than-air craft were sighted by witnesses across the western, southwestern, and central part of the United States. Although there were some allegations at the time that these airships were nothing other than experimental types of dirigibles, witnesses reported ornate designs and advanced technology. Part of the myth has been that these airships sprang from the minds of a group of air enthusiasts in the American southwest, and of particular interest to UFO researcher Jim Marrs is a well-documented crash in Aurora, Texas.

Greenewald, John, Jr.

Founder and webmaster of the "Black Vault" Internet site, one of the most important and well-researched websites for UFO-related documents. Greenewald has collected endless pages of government documents that validate the UFO phenomenon.

Greer, Steven

Steven Greer

An emergency room physician whose Center for the Study of Extraterrestrial Intelligence (CSETI) and Disclosure Project have become two important endeavors in UFO research. Dr. Greer, who claimed to have had an out-of-body experience when he was a young man during which he encountered extraterrestrials, believes that the aliens visiting earth are benevolent and that for economic and purely self-serving reasons, the government is not only withholding facts about its dealings with them, but also withholding facts about ET technologies. *See* **Center for the Study of Extraterrestrial Intelligence (CSETI)**

Groom Lake

See Area 51; Dreamland; S-4

Grudge

See Projects, Government UFO

Gulf Breeze Incident

The Gulf Breeze sightings of 1987 were UFO encounters that centered around a small residential community situated near a series of huge U.S. military complexes in Pensacola, Florida. Building contractor/construction manager Ed Walters was the first to see strange lights floating overhead. By the time the sightings ended five months later, Walters had taken scores of photographs, some with a simple Polaroid camera, others with specially sealed and preloaded cameras supplied to him by MUFON. Walters also claimed to have experienced physical encounters with alien creatures. The Gulf Breeze sightings made national news headlines.

Debunkers and skeptics decried Walters' stories as pure fabrication and held every aspect of his life up to a disturbing scrutiny. Evidence was also presented purporting to prove that Walters had faked his UFO photographs, photographs which MUFON's Walt Andrus himself seemed to have validated.

Even after the national controversy died down, sightings continued around Gulf Breeze, this time of flying triangles similar to those sighted over Phoenix, Arizona in 1997. Allegations against Walters also continued, but whether these stories were true or simply fabrications, they shed no actual light on the veracity of the sightings themselves. Gulf Breeze still remains a mystery. *See* **Andrus, Walter; Flying Triangles; Pensacola, Florida; Walters, Ed**

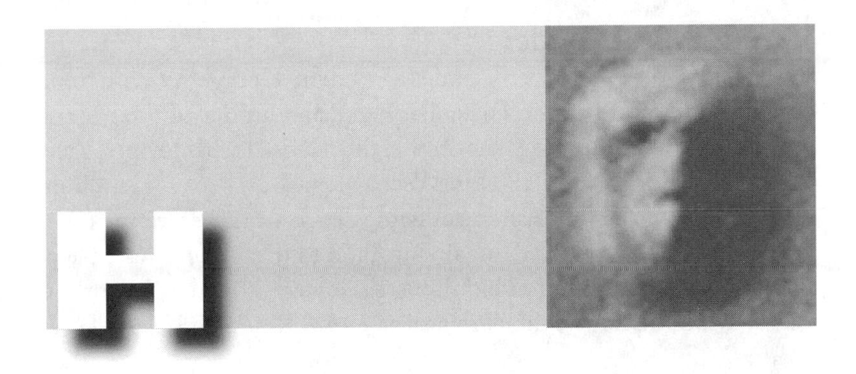

HAARP

See **High Frequency Active Auroral Research Project**

Haines, Gerald

Historian who prepared a report for the CIA publication *Studies in Intelligence* while working at the National Reconnaissance Office in 1997. *See* **Central Intelligence Agency, History of UFOs in; Disinformation**

Hall, Richard

UFO Magazine columnist, researcher, and archivist, Hall worked with Maj. Donald E. Keyhoe as his top aide at the National Investigations Committee on Aerial Phenomena (NICAP) in Washington, D.C. Hall served as NICAP's executive secretary and assistant director and now manages the Keyhoe Archives. *See* **Keyhoe, Donald; National Investigations Committee on Aerial Phenomena (NICAP)**

Halt, Charles

U.S. Air Force Lt. Col. who was the deputy commander at RAF Bentwaters in 1980 and reportedly one of the high-ranking officers who encountered a UFO in Rendlesham Forest and communicated with the alien creatures aboard. Lt. Col. Halt's report of the mysterious lights, along with the release of selected parts of the Bentwaters file, caused a sensation and much controversy within the UFO research community. *See* **Bentwaters Case**

Hamilton, Alexander

Secretary of the Treasury under President George Washington, one of the founders of the Federalist Party, and author of the Federalist Papers, Hamilton was shot to death in a duel with former Vice President Aaron Burr in 1804 in Weehawken, New Jersey. Hamilton was also one of the first American witnesses to what, before the time of flying saucers, was called an "airship," and to what might have been the first cattle mutilation. Hamilton's encounter is related by author Linda Moulton Howe in *Glimpses of Other Realities, Vol. II, High Strangeness*:

> Hamilton was alerted to the sound of the airship by the sound of loud humming, and he watched it slowly descend over his cattle yard until it hovered just overhead. Hamilton wrote:
>
> "It consisted of a great cigar-shaped portion possibly three hundred feet long, with a carriage underneath. The carriage was made of glass or some other transparent substance alternating with a narrow strip of some material. It was brilliantly lighted within and everything was plainly visible. It was occupied by six of the strangest beings I ever saw. They were jabbering together, but we could not understand a word they said. It seemed to pause and hover directly over a two-year-old heifer, which was bawling and jumping, apparently fast in the fence. Going to her, we found a cable about a half-inch in thickness made of some red material, fastened in a slip knot around her neck, one end passing up to the vessel, and the heifer tangled in the wire fence. We tried to get it off but could not, so we cut the wire loose and stood in amazement to see the ship, heifer and all, rise, slowly, disappearing in the northwest. Neighbor Thomas Link (four miles away) found the hide, legs, and head in his field the next day and no tracks in the soft ground, which mystified him."

Hamilton, William

Executive director of Skywatch International, Hamilton is an information technologist and UFO researcher who served a tour of duty with the U.S. Air Force Security Service from 1961 to 1965 as an electronic technologist. A member of MUFON, Hamilton has written *Cosmic Top Secret*, *Alien Magic*, and *The Phoenix Lights*.

William Hamilton

Currently a senior programmer-analyst at UCLA, Hamilton has investigated the Phoenix Lights sightings, Area 51, "Stargate" phenomena, alien technology and biotechnology, and the Antelope Valley sightings in California.

Hangar 18

The facility at Wright-Patterson Air Force Base in Ohio where, reportedly, the Roswell alien bodies were stored in 1947. Wright-Pat was home to the U.S. Air Force's Foreign Technology Division, where recovered Russian and Eastern Bloc high-tech components were examined, and thus it would have been an obvious place to keep saucer crash debris and alien remains, if any existed. A cable network documentary film crew was granted permission in the late 1990s to photograph the "hangar," which, by then, had been turned into a jet-engine testing cell. As the film footage showed, there were several curious-looking storage compartments built into the testing chamber's walls, covered by ancient-looking heavy iron doors—perhaps the remains of decades-old freezer compartments used for storing something really "foreign"?

Hangar 84

The hangar at the Roswell Army Air Field where, reportedly, the material from the crashed Roswell saucer and the bodies of its alien passengers were taken before they were flown to Fort Worth, Texas.

Hanger 84 at the Roswell Army Air Field in July, 1997.

Harzan, Jan C.

MUFON California State Section Director who, at a reception for Ben Rich, former head of Lockheed Aircraft and the author of *Skunk Works*, was told by Rich that the U.S. "already has the technology to send ET home." Rich complained that even though the United States had the technology to allow it to travel among the stars, that technology was being kept from the rest of humanity. Harzan said that those in the small group to whom Rich was speaking were absolutely stunned not only by what he said, but by the implications that humans are capable of star travel and that there actually are ETs to send home. *See* **Rich, Ben; Skunk Works**

Haut, Walter

The Public Information Officer for the 509th Bomber Squadron at the Roswell Army Air Field in 1947 and, in his own words, an unofficial aide to Col. Thomas Blanchard, base commander. Lt. Haut was ordered by Col. Blanchard to write the now-famous press release revealing to the public that a retrieval team from the 509th had recovered the debris of a flying saucer that had crashed in the desert outside of Roswell, New Mexico. *See* **Children of Roswell; Roswell, New Mexico; Roswell Crash Press Release**

Hesemann, Michael

Germany's most prolific UFO researcher; co-author of *Beyond Roswell,* who has conducted extensive interviews with officers and managers of the former Soviet space program, Hesemann has gotten highly placed Soviet officials to admit on camera that not only were the scientists and military officers in the Russian space program aware of the existence of UFOs, but that aliens are interacting with human beings. *See* **Russian Roswell**

Michael Hesemann

Hickson, Charles

Mississippi shipyard worker who was abducted, along with Calvin Parker, in a 1973 incident. *See* **Parker, Calvin; Pascagoula Affair**

Hieronimus, Robert R.

Artist, author, and radio broadcaster who has been a frequent guest on radio and television talk shows across the country since 1967. Bob Hieronimus' "21st Century Radio" program is broadcast on over a hundred stations nationwide on the American Radio Network. He is the bestselling author of *How to Pick Your Personal Winning Lottery Numbers,* and *Your Personal Winning Lottery Numbers*. His most recent book is *Inside the Yellow Submarine: The Making of the Beatles Animated Classic.*

Hieronimus, Zoh

Baltimore native Zohara Hieronimus is the radio talk show host of "Future Talk," currently syndicated nationally and featuring experts from around the world in fields of science, healing, global affairs. She is married to Dr. Robert Hieronimus, with whom she has co-hosted cable and radio programs, including "21st Century Radio."

High Frequency Active Auroral Research Project (HAARP)

According to its stated objective, HAARP (the High Frequency Active Auroral Research Project) begun in 1991, is an American scientific research program aimed at studying the effects of high-energy electrical transmissions on the ionosphere, that part of the earth's atmosphere fifty to four hundred miles in altitude, containing electrically-charged particles. Opponents of HAARP maintain that these powerful transmissions, useful for civilian and military communication and navigation systems, are ultimately harmful to the fragile ecosystem which supports the planet, and that HAARP represents a technology that could ultimately lead to a new class of weaponry that would profoundly upset the equally fragile geopolitical power system.

Hill, Barney and Betty

This New England couple became the first abductees of modern record. On September 19, 1961 Betty and Barney Hill were driving home at night along a country road in rural New Hampshire when they realized they were being tracked by a bright light in the sky. Barney eventually stopped the car when the bright light appeared on the road in front of them. The next thing they knew they were

Betty and Barney Hill, two of the earliest known abductees, at their Maine home in the 1960s.

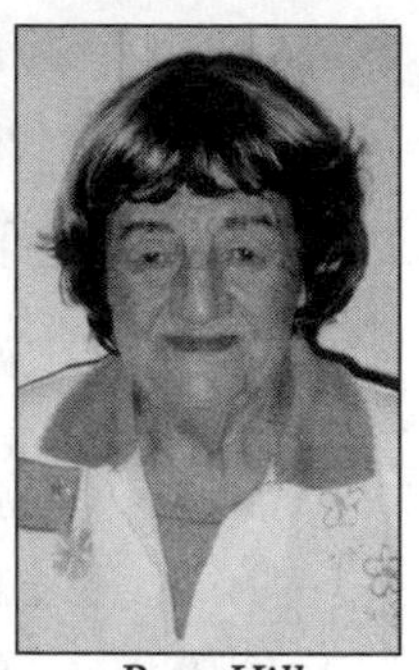
Betty Hill

back on the road heading for home, but it was two hours later and they couldn't account for the missing time.

Months later, after Barney's continued bouts of anxiety and physical symptoms, the couple sought medical attention. Their doctor could find no underlying medical problems and recommended they visit a psychiatrist, Dr. Benjamin Simon. Dr. Simon regressed both Betty and Barney separately and recorded their stories of an abduction by strange beings who took them aboard a spaceship.

Betty Hill believes that Dr. Simon's assistant, who had access to the notes the psychiatrist made, delivered the Hills' case file to a writer for the *Boston Globe*, who printed the story in the newspaper. That was how their abduction became public knowledge. It's important to note that the Hills were a racially mixed couple who were private people with no intentions of publicizing their experience. After several decades, few, if any, believe their story is anything but true.

One of the more remarkable aspects of the Hill case is the "star map" that Betty was able to sketch while under hypnosis, reproducing in detail the map her alien abductors had shown her aboard their spacecraft. The map, she was told, showed nearby inhabited planets, and astronomers were later able to infer from the detailed sketch she made where the aliens' home world was located.

Barney Hill died in 1969. After his death, Betty was approached by actor James Earl Jones, who was fascinated with the story of the abduction. Jones produced a movie for television in which he played Barney Hill and Estelle Parsons played Betty. It was one of the first docu-dramas about a supposed real UFO encounter. Betty Hill has continued to speak and write about UFOs, her abduction experience, and the nature of the extraterrestrials who continue to visit earth. See **Fish, Marjorie**

Hill, Paul

Author of *Unconventional Flying Objects,* Hill was himself a witness to a 1952 incident in which he saw a group of four UFOs organize themselves into a V-formation in the skies over Chesapeake Bay. Hill's book is a scientific analysis of the kinds of craft and propulsion systems that might power them. Among his conclusions were that conventional propulsion systems were inadequate to power interstellar craft and that UFO engineers had managed to develop a "field engine."

"The essence of the field engine," he wrote, "is a static field link between the UFO and the earth or other large mass, planetary or stellar." Hill's prediction as to the nature of a field engine prefigured research into zero-point energy and the revelations of Robert Lazar and others about anti-gravity and magnetic field generator experiments being conducted at Area 51. *See* **Area 51; Lazar, Robert**

Paul Hill

Hillenkoetter, Roscoe

The first Director of the CIA (1947–1950), shepherding it from its incarnation as "central intelligence" to a full-fledged bureaucratic agency. Later, Adm. Hillenkoetter became a member of the National Investigations Committee on Aerial Phenomena (NICAP). According to Stanton Friedman, Adm. Hillenkoetter was a member of MJ-12 and unequivocal in his belief in the existence of UFOs. He wrote:

> It's time for the truth to be brought out in open congressional hearings. Behind the scenes, high-ranking Air Force officers are soberly concerned about UFOs. But through official secrecy and ridicule, many citizens are led to believe the unknown flying objects are nonsense. To hide the facts, the Air Force has silenced its personnel through the issuance of a regulation.

See **Majestic; National Investigations Committee on Aerial Phenomena (NICAP)**

Hoagland, Richard

A former NASA and CBS News consultant during the Apollo moon missions, Richard Hoagland was one of the originators of the "calling card" plaque that was placed onboard the Pioneer 10 space probe set to leave our solar system, announcing the presence of human beings and their attempts to make contact with other species.

Richard T. Hoagland with a photograph of the "Face."

Hoagland has since become a

critic of NASA, arguing that the agency is withholding from the American people evidence it has of a former civilization on Mars. The evidence supports his hypothesis that the "Face on Mars" and its accompanying "pyramids" are constructs of an alien civilization and not the result of natural planetary forces.

Though often a thorn in the side of the scientific establishment, who publicly deride him, Hoagland has found support for many of his theories from NASA and JPL scientists and engineers, who often surreptitiously provide him evidence for his ongoing investigations. Hoagland is the author of several books, including *The Monuments of Mars*. *See* **Mars, Face on**

Holden, Bill

Steward aboard Air Force One, to whom President Kennedy once revealed that he would like to "tell the public about the alien situation, but my hands are tied."

Holden, W. Curry

An archaeologist from Texas Tech University. Holden was reportedly on a field trip with a group of students in 1947 when they came upon the Corona (Roswell) crash site. Thinking that an airplane had crashed, Holden sent one of his students to call the local sheriff. While waiting, the group saw three small bodies lying near the craft. When the military arrived, Holden and his students were escorted to the Roswell Army Air Base, where they were detained and warned against talking about what they'd seen.

Later that same day in Roswell, Holden ran into another scientist, Dr. C. Bertrand Schultz, a paleontologist who had been working in the area. Holden told Schultz what he had seen, and Schultz was impressed enough that he recorded the meeting with Holden in his personal diary. Holden later confirmed in an interview that he had in fact been at the site of the crash. *See* **Roswell, New Mexico; Schultz, C. Bertrand**

Hoover, George W.

One of two officers at the Office of Naval Research who read and no doubt commented upon the annotated version of Morris K. Jessup's *Case for UFOs,* along with Capt. Sidney Sherby. Commander George W. Hoover privately financed a mimeographed edition of the book. Hoover was one of the Navy's most innovative thinkers and his concept for a flat-panel display led to the development of today's overhead "heads-up" display in cockpits. In fact, the single copy of the annotated Jessup book remained in Hoover's possession until he died, along with one of the copies of the "Varo Edition" of the book, named after the company that did the printing. *See* **Jessup, Morris K.; Philadelphia Experiment**

Hoover, J. Edgar

The founding director of the Federal Bureau of Investigation, reported to be at the center of more conspiracies than any other senior member of the government. According to FBI memos released under Freedom of Information Act requests, Hoover was interested in finding as much information as he could about UFOs, but his inquiries were consistently blocked by the military. Privy to many, but not all, of the UFO secrets being passed around at the highest levels of government at the time, and no doubt intrigued by their import, Hoover, obviously miffed at being kept "out of the loop," wrote in a 1947 memo that he would only investigate flying saucer incidents if the military cooperated with the FBI and granted them access to the information they had already assembled. *See* **Federal Bureau of Investigation (FBI)**

Hopkins, Budd

Budd Hopkins

UFO abductee researcher and author Hopkins was among the first to conduct ongoing in-depth research into the phenomenon of missing time and abduction experiences. Hopkins' personal experiences with those who have had close encounters and suffered missing time episodes led him into deeper involvement in the study of the abduction phenomenon, and he's since conducted hundreds of regression-therapy sessions with abductees. His books on the subject include *Missing Time, Intruders: The Incredible Visitations at Copley Woods,* and *Witnessed,* the last, in which Hopkins writes about the abduction of Linda Cortile from her apartment in Manhattan. Hopkins has become one of the most important alien abduction researchers along with John Mack and David Jacobs. *See* **Brooklyn Bridge Abduction Case; Copley Woods**

Hopkins, Herbert

A therapist in Maine who might have been involved in one of the earliest known "men in black" cases. According to Peter Brookesmith's book *UFO: The Complete Sightings*, Dr. Hopkins had been treating a young abduction patient who also claimed to have been visited by a strange, menacing individual who warned him never to talk about UFOs or his experiences. Hopkins then received a strange visit, himself, from a hairless man with a sickly pale complexion, who was wearing lipstick(!) and dressed all in black. Another bizarre episode followed a short time later when Dr. Hopkins' son and daughter-in-law received a visit from a strange couple also exhibiting abnormal behavior.

Some UFO researchers believe that these visitors might have been extraterrestrials and not human beings working for the government, leading to the speculation that the aliens themselves might be men in black. *See* **Men in Black**

Howe, Linda Moulton

Linda Moulton Howe

Author, Emmy-award-winning documentary filmmaker, and researcher who was one of the first to investigate and write about the phenomenon of cattle mutilations. In her documentary *A Strange Harvest*, Howe connected cattle mutilations to two main sources: UFOs, whose appearances in areas of mutilations are well documented; and what have been called "black helicopters," secret military helicopters operated by a branch of the U.S. government. Howe has written that government intelligence operatives have told her that cattle mutilations are carried out by aliens to test the level of toxins in earth's environment. And the black helicopters are a deliberate attempt by the government to divert attention from UFO sightings. *See* **Black Helicopters; Cattle Mutilations**

Hubbell, Webster

President Clinton's number-two man at the Department of Justice, who in 1993 was asked by Clinton "to find the answers to two questions for me: Who killed JFK and are there UFOs?" Hubbell said that he was completely stonewalled in his attempts to find the answers, and President Clinton was thereby denied any access to UFO secrets. *See* **Clinton, William Jefferson**

Hubble Space Telescope

An orbital space telescope, which, when launched, held the promise of obtaining high-resolution images of objects in space unobstructed by the earth's atmosphere. The Hubble has indeed taken stunning images of objects that are millions of light-years away, yet NASA claims the telescope can't produce decent images of the moon, which is only 250,000 miles distant. For subscribers to the "Moongate" theory, NASA's refusal to release these images indicates that it's trying to hide the truth about artificial structures on the lunar surface. *See* **Moon, Dark Side of the**

Hudson Valley Overflights

J. Allen Hynek, Philip Imbrogno, and Bob Pratt described the numerous sightings in the late 1970s over New York's Hudson Valley that were witnessed by thousands of local residents in their book *Night Siege: The Hudson Valley UFO Sightings. See* **Flying Triangles**

Human Mutilations

Either a parallel to the now well-documented cattle mutilations or a separate set of incidents, human mutilations involve stories of people whose bodies have been found drained of blood, with their sex organs removed, and other soft tissue excised as well. Stories of human mutilation experiments have involved S-4 at the Area 51 base and those who have claimed to have first-hand information have said that humans are the subjects of gruesome experiments in hybridization. It is claimed that because of genetic experiments carried out in the future by creatures who are not extraterrestrials but human survivors of some great tragedy, time-traveling visitors must return to our present to refresh their human DNA. These creatures, in league with governments of the world, experiment on humans for their own purposes, and in return, share future technology with friendly governments. One of the most prominent exponents of human mutilation stories is John Lear. *See* **Cattle Mutilations; Lear, John**

Hunsaker, Jerome

Chairman of the departments of mechanical and aeronautical engineering at MIT; member of the MJ-12 committee. *See* **Majestic**

Hynek, J. Allen

One of the most influential figures in the history of UFO research. Hynek's conversion from UFO skeptic to believer is one of the enduring stories of ufology. Chairman of the Astronomy Department at Chicago's Northwestern University, Hynek was engaged by the U.S. Air Force in 1948 to work for Project Sign as an evaluator of witness reports of UFO sightings. Hynek went on to consult for Project Grudge and Project Blue Book.

Dr. J. Allen Hynek, a confirmed skeptic who would later become a qualified believer.

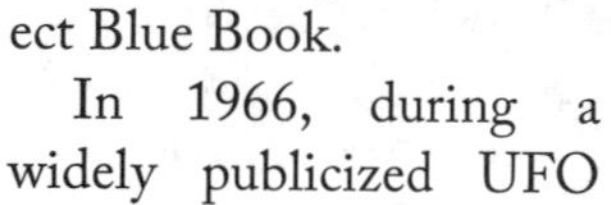

In 1966, during a widely publicized UFO flap in Michigan, Hynek made the public statement that he believed some of the objects being observed were merely "swamp gas." That statement would come to haunt him, as the astronomer was flayed by UFO skeptics and believers alike. By the time Blue Book was shut down after the Condon Report in 1969, Hynek had become disillusioned with the Air Force's continued stonewalling over the UFO matter and by the preponderance of the affirmative evidence he'd seen over the years. He went on to become one of the founders of the Center for UFO Studies (CUFOS).

Dr. Hynek, who died in 1986, reportedly said, "When the long-awaited solution to the UFO problem comes, I believe that it will prove to be not merely the next small step in the march of science but a mighty and totally unexpected quantum jump." *See* **Center for UFO Studies (CUFOS); "Invisible College"**

Hynek Classification System

A system of classifying UFO sightings developed by Dr. J. Allen Hynek in the 1950s while he was doing research on the subject for the Air Force:

Nocturnal Light

Any anomalous lights) seen in the night sky whose description rules out the possibilities of aircraft lights, stars, meteors, and the like.

Daylight Disk UFOs

Seen in the distant daytime sky. The UFOs classed in this

category can be other shapes, as well, like cigars, eggs, and ovals.

Radar Visuals

Where UFOs are tracked on radar and can be seen at the place illustrated at the same time. A good example would be a disk-shaped object traveling at 9,000 miles per hour and showing up on radar in 1950 when the world air-speed record was only 650 miles per hour.

Close Encounters of the First Kind

A UFO in close proximity (within approximately 500 feet) to the witness.

Close Encounters of the Second Kind

A UFO that leaves markings on the ground, causes burns or paralysis to humans, frightens animals, or interferes with car engines or TV and radio reception.

Close Encounters of the Third Kind

A UFO which has visible occupants.

Two other classifications later added to Hynek's original list are:

Close Encounters of the Fourth Kind

These include alien abduction cases.

Close Encounters of the Fifth Kind

Where communication occurs between a human and an alien being.

Hypnosis, Abduction Memory Recovery

In general, and specifically as it applies to the technique used to recover memories of alien abduction, no method has aroused more controversy than the use of hypnosis. Particularly because hypnosis is described as an ultra high state of suggestibility in which the subject's conscious, logical, and critical awareness is focused away from censoring information stored in the subject's subconscious, the subject is absolutely vulnerable to the influence of the therapist even when the therapist is unaware that he or she is exerting any influence upon what the subject is reporting. Accordingly, testimony given under hypnosis is only admitted into evidence in court under very tightly controlled conditions, if it is admitted at all, and testimony elicited via hypnosis is itself highly suspect.

As a basis for recovering memories, hypnosis is also useful, though highly suspect, because of the influence of the person guiding the subject through memories. Because of the work of Dr. Elizabeth Loftus on False Memory Syndrome and the political overhead that FMS has had in professional psychology circles, hypnotic regression that is used to recover memories of alien abduction remains a highly charged issue.

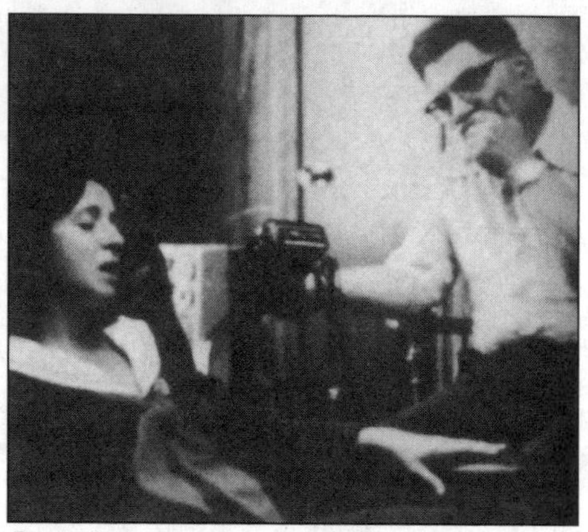

Betty Andreasson Luca under hypnosis talking to researcher Joseph Santangelo.

Skeptics of abduction therapy have demonstrated that subjects under hypnosis, without any affirmative or deliberate implanting of memories, can be made to create an abduction scenario out of whole cloth because 1) the subject knows what the therapist wants to hear, and 2) hypnotized subjects are very imaginative in creating fictional events which they later believe to be real, especially when alien abductions are extensively featured in science-fiction books and movies, and the news.

But none of this can discount the testimony of witnesses such as Barney and Betty Hill, who underwent therapy at a time before abduction stories became the province of the media. Moreover, the work of psychiatrist John Mack at Harvard and Professor David Jacobs in Pennsylvania also indicates that something is going on the minds of people reporting these experiences that has not been put there by a regression therapist. Budd Hopkins has provided additional validation through his compilation and categorization of the types of abduction experiences and the entire concept of missing time common to the overwhelming majority of abductees. *See* **Abductees; Hopkins, Budd; Jacobs, David; Mack, John**

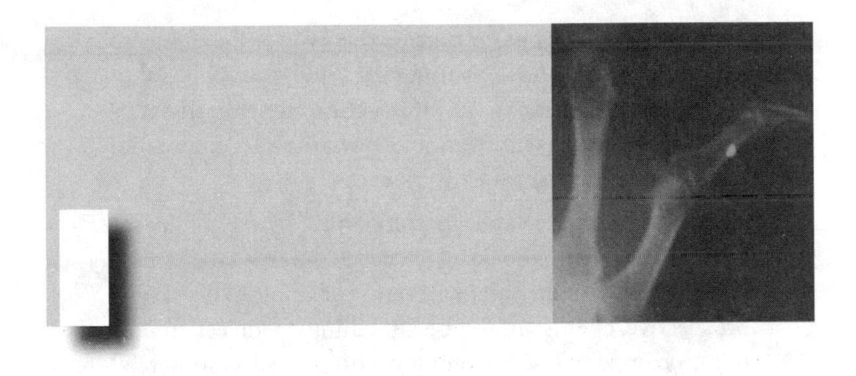

Icke, David

Author, speaker, and political activist, David Icke is a former national spokesperson for the British Green Party and he has become one of the most controversial personalities in ufology. His theories, a blend of spiritualism, psychic phenomena, political and exobiological conspiracy, and historical revisionism, have made him a popular speaker on issues of the day. Perhaps his most controversial theory is that the world leaders, including the royal family and a number of U.S. presidents, including George Bush, are not actually human beings but reptilian extraterrestrials who, since Biblical times, have been enslaving humanity and guiding the course of events on planet earth for their own benefit. *See* **Alien Beings; Reptilians**

Ikonos I Satellite

On April 27, 1999, an Ikonos 1 commercial imaging satellite launched from Vandenberg Air Force Base disappeared before it reached its intended earth orbit. Later investigation revealed that the satellite hadn't exploded or fallen back into the atmosphere, but the ultimate cause of the failure was "undetermined." While UFO researchers have suggested that one or more extraterrestrial groups are interfering with our space program, those in the military and commercial aerospace community offer other explanations:

1. Companies that launch commercial surveillance satellites that could photograph American military installations

have been put on notice that in the event of a threat to national security, those commercial satellites will be "shut down" by our military. The destruction of selected satellites is a test of that system and a warning to companies that the U.S. military means business.

2. Rival commercial satellite companies, vying for lucrative contracts, are knocking out each other's satellites. If a company's satellite fails without reasonable explanation, prospective clients might be unwilling to invest money in the company's subsequent ventures, and commercial insurers will not cover their products.
3. U.S. adversaries, such as Russia or China, are deploying weapons that have the capability of knocking out our satellites. Directed-energy weapons, such as lasers and particle beams, are being developed by several nations, and their "death rays" usually leave no trace of a targeted satellite.

Implants, Alien

Odd-looking metallic or ceramic devices that many abductees claim have been implanted beneath their skins during medical tests performed on them by alien abductors. While these claims have been considered paranoid delusions by almost all skeptics and even some believers, beliefs changed dramatically when Dr. Roger Leir, a Los Angeles podiatric surgeon, went public with reports of operations he'd performed on implantees and the nature of the devices he'd extracted.

Some implants are small and round, the size of a BB, while others are no larger than a pinhead. They come in many different shapes: round, triangular, spiral, and thin as a fine wire. The implants are clearly visible on the hun-

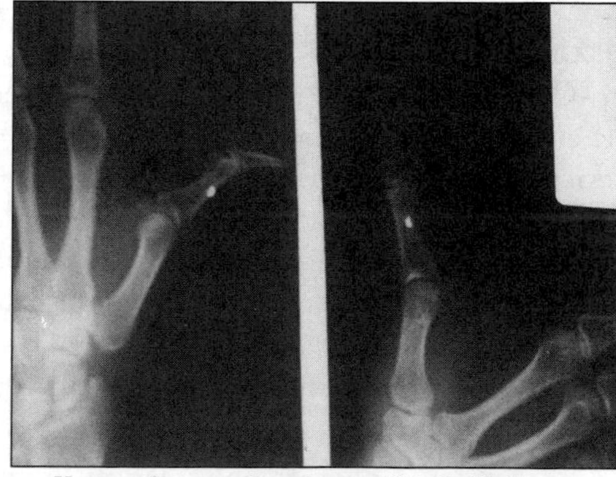

X-rays of suspected implants still imbedded in an abductee's thumb.

dreds of X-rays that have been taken and documented by researchers. Most of the implants found so far have been placed on the left side of an individual's body.

While in place, these implants are surrounded by an unusually strong covering of living tissue comprised of protein and keratin, the elements comprising human hair and fingernails. This covering is so tough that surgeons cannot cut it with a hardened steel scalpel. There is no substance on earth made by man that can match the strength of this simple outer covering.

Three researchers in the troubling area of alien abduction and implantation look at an actual implant. Left to right: Dr. Roger Leir, author Whitley Strieber, therapist Derrel Sims.

Prior to being removed, implants can often produce a very strong electromagnetic field. A Gauss meter registers off-scale when it's held next to the area of the body where the implant is located. However, once an implant is removed, it no longer registers on the meter, and it is then nearly impossible to cut through the tough outer cutaneous-like covering of the device. The implication is that a person's own energy field is powering the implant, allowing it to generate the force field.

While the implant is in place, there are large bundles of sensitive nerves surrounding and growing into it; specifically, the same types of nerves that allow a person's hand to hold a glass with the right amount of pressure so as not to crush it or drop it. Implants are often found in places where these kind of nerves don't normally grow, such as the back of the hand or in the jaw.

These devices have been examined by some of the world's leading metallurgy labs, including the federal government's own national facility at Los Alamos, New Mexico. The Los Alamos lab concluded that the implants must be pieces of a meteorite, because no such metallic isotopes are found in such proportions on earth.

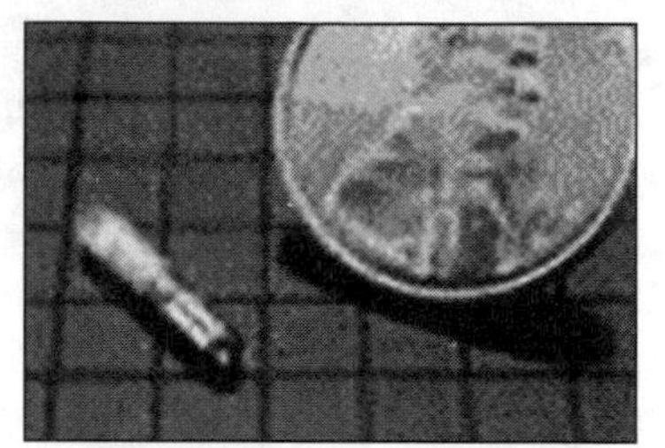

Actual chip being considered by industry (terrestrial) for human implantation.

Reactions of the abductees themselves to the implanted devices have been remarkable, according to abduction therapists. While most abductees believed the implants were like bar-coded identification strips such as those found on products in stores to be scanned at the checkout counter, others believe them to be GPS locator devices like those used to track stolen automobiles. Abductees have reported that once the devices had been removed, they felt a sense of loss, as if part of their personalities had been excised. This has led some researchers to believe that implants also serve as interactive surveillance devices, constantly sending and receiving information from their extraterrestrial monitors. *See* **Leir, Roger**

"Infancy and Protevangelion," Books of

These "lost books" of the Bible report many incidents of UFO encounters and abductions. In one especially vivid narrative, Joseph describes an incident that happened to him just before the birth of Christ:

> I looked into the air, and I saw the clouds astonished, and the fowls of the air stopping in the midst of their flight. And I looked down towards the earth, and saw a table spread, and working people sitting around it, but their hands were upon the table, and they did not move to eat. They who had meat in their mouths did not eat. They who lifted their hands up to their heads did not draw them back. And they who lifted them up to their mouths did not put anything in, but all their faces were fixed upwards. And I beheld the sheep dispersed, and yet the sheep stood still. And the shepherd lifted up his hand to smite them, and his hand continued up. And I looked unto a river, and saw the kids with their mouths close to the water, and touching it, but they did not drink.

Then a bright cloud overshadowed the cave, and the mid-wife said, "This day my soul is magnified, for mine eyes have seen surprising things, and salvation is brought forth to Israel." But on a sudden the cloud became a bright light in the cave, so that their eyes could not bear it. And, behold, it was all filled with lights, greater than the light of lamps and candles, and greater than the light of the sun itself.

Inman, Bobby Ray

A former Naval intelligence officer and one-time deputy director of the CIA who, some UFO researchers believe, has been at or close to the center of secret UFO programs. Adm. Inman himself has suggested that there is more to the UFO question than the government is willing to release.

International Flying Saucer Bureau

Founded by Albert K. Bender, editor of *Space Review*, this organization collected data and information on UFO sightings and incidents.

"Invisible College"

Astronomer J. Allen Hynek's description of a group of scientists and astronomers working on the subject of UFOs who preferred not to be identified so as to remain untainted by the stigma of frowned-upon UFO research, even though that research may have been conducted at the behest of the military or the government.

Ishikawa, Kanshi

In a 1974 article in *UFO News,* General Ishikawa, the chief of staff of Japan's self-defense force said, "Much evidence tells us UFOs have been tracked by radar; so UFOs are real and they may come from outer space ... UFO photographs and various materials show scientifically that there are more advanced people piloting the saucers and motherships."

Iowa Family Abduction Case

In a case reported by investigators Beverly Trout, Irene Barnes, Lawrence Lacey, and Robert Lyon in the January 1998 issue of the *MUFON Journal*, an entire family in Iowa, whose names were not released, had an abduction experience while driving to visit a relative.

In 1997, as "Bob" was driving his eleven-year-old nephew to visit

Bob's mother, they saw an amber light shining through the trees of Stone State Park, a light that appeared to be moving in the same direction and at the same speed that they were. They could see the object was triangular in shape and quite large, about seventy feet in length, with three lights defining its perimeter. The object then banked toward them, coming so close that they could clearly see the details of its underside.

Afterwards, just as Bob and his nephew turned off the main road, "something" ran into their car, hitting hard against the side. They looked out the window and saw a muscular, fur-covered animal with a hump on its back. It was snarling, flashing huge teeth and blood-red eyes. Bob didn't wait around to get a closer look, but floored the car's accelerator to get away.

After arriving at Bob's mother's house, all three members of the family saw an orange globe floating in the sky about a mile away. The globe pulsated and then split into two smaller globes. The family continued to watch this display for twenty minutes.

Later that night Bob's young nephew had a strange nightmare in which he became paralyzed and felt a needle being stuck into his arm. Three days later the grandmother noticed unusual puncture wounds on the boy's fingers and toes. Several weeks after that, the grandmother and some of her neighbors observed strange lights flying around the sky, close to her home. The lights reappeared over the next several nights.

The following month Bob and his sister were out driving when they saw a silver flying disk approach them. They panicked and drove off. A week later a large cigar-shaped craft flew overhead, and Bob could clearly distinguish several rectangular windows on the craft. He also felt as if he were receiving communications from the beings aboard the craft. These series of events left Bob and his family feeling fairly certain they have been abducted on more than one occasion.

Jacobs, David

A professor of history at Temple University in Philadelphia. David Jacobs' 1974 doctoral dissertation was on the phenomenon of UFOs in America. He ultimately published a book based on his doctoral research entitled *The UFO Controversy in America.* In 1992 he published *Secret Life: Firsthand Accounts of UFO Abductions.* Jacobs has become one of the foremost researchers investigating UFO abductions in America along with Dr. John Mack and Budd Hopkins.

Professor Jacobs has discovered that abductees come from all walks of life, from every income level, and include all races, religions, and cultures from all areas of the country. Jacobs has concluded that the aliens responsible for these abductions are neither passive nor benevolent. He has said that we are their lab rats, and our resources might be theirs for the taking. He presents a dark scenario regarding the intentions of hostile extraterrestrials who visit our planet and help themselves to human test subjects. *See* **Abductees**

Jarod 2

An anonymous seventy-year-old former Area 51 worker nicknamed "Jarod 2," who confirms Robert Lazar's story that aliens were held at the secret complex. One of those beings, known as "Jarod," was kept alive and housed at the base. Jarod 2 says he worked on designing reproductions of alien flying saucers and that he has actually seen the gray aliens living at the base. *See* **Area 51; Lazar, Robert**

Jessup, Morris K.

The genesis of the Philadelphia Experiment myth dates back to 1955 with the publication of *The Case for the UFO* by the late Morris K. Jessup.

Some time after the publication of the book, Jessup received correspondence from a Carlos Miguel Allende, who gave his address as R.D. #1, Box 223, New Kensington, Pennsylvania. In his correspondence, Allende commented on Jessup's book and gave details of an alleged secret naval experiment conducted by the Navy in Philadelphia in 1943. During the experiment, according to Allende, a ship was rendered invisible and teleported to and from Norfolk in a few minutes, with some terrible after-effects for crew members.

This is the actual cover of the Jessup book, as annotated by Naval commander George Hoover.

Supposedly, this incredible feat was accomplished by applying Einstein's unified field theory. Allende claimed that he had witnessed the experiment from another ship and that the incident was reported in a Philadelphia newspaper. The identity of the newspaper has never been established. Similarly, the identity of Allende is unknown, and no information exists on his present address.

In 1956 a copy of Jessup's book was mailed anonymously to the Office of Naval Research. The pages of the book were interspersed with handwritten comments which alleged a knowledge of UFOs, their means of motion. Allende described the culture and ethos of the beings occupying these UFOs in pseudo-scientific and somewhat incomprehensible terms.

Two officers, then assigned to ONR, took a personal interest in the book and showed it to Jessup, who concluded that the writer of the comments in his book was the same person who had written him about the Philadelphia Experiment. These two officers personally had the book retyped and arranged for the reprint, in typewritten form, of twenty-five copies. The officers and their personal belongings left ONR many years ago, and ONR does not have a file copy of the annotated book. *See* **Hoover, George W.; Philadelphia Experiment**

Jet Propulsion Laboratory (JPL)

Managed by the California Institute of Technology, JPL is NASA's center for exploration of the solar system. JPL spacecraft have visited all the planets in the solar system except Pluto, and their telescopes are observing distant galaxies to study how our solar system was formed. JPL also manages the worldwide Deep Space Network, a collection of huge tracking antennas used to communicate with spacecraft, and conducts scientific investigations from its facilities in California's Mojave Desert near Goldstone; as well as Madrid, Spain; and Canberra, Australia.

Mars Polar Lander, according to an artist's conception. The project team from JPL lost contact with the lander on December 3, 1999.

According to some researchers, JPL is responsible for withholding information from the public on what it has really learned from its space probes. From the "Face on Mars" to possible life on other planets and especially with respect to UFO sightings, JPL, some researchers say, operates to filter out information that NASA does not want released. Most notably the 1999 failures of its two Mars missions are examples of the way JPL covers up the fact that something out there is interfering with our space probes. *See* **Mars; National Aeronautics and Space Administration (NASA)**

Johnson, J. Bond

The photographer who took the famous photo of Gen. Roger Ramey kneeling over the Roswell wreckage with a telegram in his hand. Johnson was a young reporter for the *Fort Worth Star Telegram* in 1947 when he was dispatched to Gen. Ramey's office at the Fort Worth Air Base to take some photographs. Johnson did not know the substance of the story or what he would be photographing.

Ramey had what looked like debris spread over the floor of his office, debris that carried a strong burnt smell. There was nothing recognizable about the wreckage that Johnson could identify as a weather balloon. While the material was being arranged and Johnson was taking photographs, Ramey received a phone call, then ordered one of his aides to remove the debris and bring back something from a nearby hangar. That new material was brought in, and Johnson recalls that this was definitely a weather balloon. Ramey was then photographed holding a telegram in his hand, which supposedly contained a message from higher-ups to quash the flying disk story. *See* **Ramey, Roger; Ramey Telegram**

Johnson, Kelly

One of the founders of Lockheed's "Skunk Works" division, responsible for the design of the U-2 and SR-71 spy planes. According to a *MUFON Journal* article by Bill McDonald in 1998, Johnson had direct access to the data from the Roswell spacecraft. It was this information that helped designers develop a number of secret aircraft, including the Stealth bomber, and what may be the craft that witnesses have referred to as a flying triangle. *See* **Skunk Works**

Johnson, Lyndon Baines

The thirty-sixth president of the United States, Lyndon B. Johnson remains at the center of one of this country's greatest alleged conspiracies: the assassination of his predecessor, John F. Kennedy. Johnson was Kennedy's vice president and assumed the presidency upon Kennedy's death in 1963.

President Johnson, according to the Haines Report on the CIA's History of UFOs, held a high-level meeting on UFOs in the White House in November 1964. The records of this meeting and the nature of the deliberations have never been released, fueling speculation that after the close of the Warren Commission, Johnson felt free to seek information regarding UFOs. *See* **Central Intelligence Agency, History of UFOs in; Kennedy, John F.**

Jones, James Earl

Although to many modern science-fiction fans, Jones is best known for his performance as Darth Vader in *Star Wars,* the actor also portrayed abductee Barney Hill for a television movie he produced based on the abduction story of Betty and Barney Hill. *See* **Hill, Barney and Betty**

Jorden-Kauble, Debbie

Author of the book *Abducted: The Story of the Intruders Continues,* which relates the author's lifelong experiences of being abducted by members of the military in sometimes bizarre re-creations of alien abductions. One of the subjects of Budd Hopkins' famous book *Intruders,* Jorden-Kauble describes a life that, because of the frequency of her abductions, is not her own. Her story was also the subject of the television miniseries *Intruders,* starring the late Richard Crenna.

Journal of UFO Research

In a 1981 article in this journal, Wen-Qwang wrote, "China is so vast, and UFOs are certainly being witnessed again and again throughout China." *See* **China, UFOs in**

Joyce, Frank

Manager of the local Roswell radio station in 1947. Joyce recalled a conversation he had with Lt. Walter Haut, the Roswell base information officer who personally delivered his press release about the U.S. Army's retrieval of a "flying disc." Joyce said, "Walter Haut came into the station sometime after I got his call. He handed me a news release printed on onionskin stationery, and left immediately.

"I called him back at the base and said, 'I suggest you not release this type of story that says you have a flying saucer or flying disc.' Haut replied, 'No, it's all right. I have the okay from my commanding officer.' I then sent the release on the Western Union wire to the United Press Bureau.'" *See* **Roswell, New Mexico**

JPL

See **Jet Propulsion Laboratory**

J-Rod

The invented name of the alleged Zeta alien housed at the S-4 facility at Area 51. J-Rod supposedly crashed in a UFO outside of Kingman, Arizona in 1953. Through communications with scientists at Area 51, some of whom have disclosed top secret information to UFO researchers, the Zetas are not simply extraterrestrials, but rather mutated human beings from our own future who have returned to "correct" genetic defects which plague their species. J-Rod has allegedly communicated warnings to scientists about the future of humanity and the fate awaiting planet earth within the next ten years. *See* **Area 51**

Jung, Carl G.

Swiss psychiatrist and founder of analytical psychology, Dr. Jung worked with Sigmund Freud until their formal break in 1914 after Jung's work on the subject of the unconscious disagreed with Freudian theory. Part of Jungian theory is based upon the concept of the collective unconscious in which human beings draw upon primal archetypes of image and thought. Among those archetypes is the symbol of the flying saucer, the mandala, which Jung believed to be central to the human experience.

In a 1955 article in *Flying Saucer Review*, Jung wrote that a "purely psychological explanation for flying saucers is ruled out" because the objects "show signs of intelligent guidance by quasi-human pilots." He admonishes the "authorities in possession" of information about UFOs to "enlighten the public as soon as possible."

In *Time* magazine on August 4, 1967, Jung called visual observations of UFOs by witnesses, as well as by radar observations "an established fact," and said that UFOs have also been proven through photographic evidence.

Additionally for UFO researchers, Jung's theories on the collective unconscious have helped explain the nature of the psychic superhighway that has been heavily traveled by remote viewers and those who practice astral projection and undertake out-of-body experiences. *See* **Collective Unconscious; Flying Saucers; Remote Viewing**

Kaifu, Toshiki

Japanese Prime Minister who, in a June 24, 1990, letter to Mayor Shiotani of Hakui City wrote, "I said that someone had to solve the UFO problem with far reaching vision at the same time. From the point of "people" in outer space, all human beings on earth are the same people, regardless of whether they are American, Russian, Japanese, or whoever."

Katchen, Lee

A NASA physicist who revealed the existence of the SAGE (Semi-Automatic Ground Environmental) system that NORAD uses to track all air traffic over the continental United States. Katchen claims that SAGE was becoming so cluttered with UFOs that controllers programmed the system to screen them out. According to Katchen:

> UFO sightings are now so common, the military doesn't have time to worry about them. The major defense systems have UFO filters built into them, and when a UFO appears, they simply ignore it. The filters cut out all unconventional objects or targets, because we are only interested in Russian targets, enemy targets. Something that hovers in the air and then shoots off at 5,000 miles per hour doesn't interest us because it can't be the enemy.

See **North American Air Defense Command (NORAD)**

Kaufman, Frank

An Army Counterintelligence Corps noncommissioned officer who was reassigned to Roswell as a civilian working for the military. Kaufman claims that he was on duty at a radar control station when anomalous radar targets showed up over New Mexico in July of 1947. He watched one such target "explode," and was then sent to its crash site, where he learned it was a flying saucer, and assisted with the retrieval of debris. Kaufman also claimed to have been part of the ensuing cover-up until finally going public with the "real" story years later. Kaufman's account has been called a hoax by many UFO researchers. *See* **Children of Roswell**

Kecksburg, Pennsylvania, Case

On December 5, 1965, scores of eyewitnesses observed a "fireball," which seemed to be under intelligent control, streaking through the nighttime winter sky until it crashed in the woods outside of the town of Kecksburg, Pennsylvania. Residents who had observed the object land called the Pennsylvania State Police to report a plane crash in the woods, and one witness described seeing a pulsating light coming from the area. However, the emergency phone calls also brought out a military recovery team whose presence in the area ultimately became overwhelming as they sealed off the site.

Researcher Stan Gordon interviewed local residents who reported seeing a ten-foot-tall, copper-colored saucer-like object with a gold band around its bottom and Egyptian-like hieroglyphics on it. Other witnesses watched the military load the object onto a flatbed truck and cover it with a tarp before driving it away in a large convoy. The Air Force would later claim that the object was a meteorite, while other officials speculated that it was a crashed Soviet Venus probe. To this day the Kecksburg sighting remains cloaked in mystery.

Keel, John

Author and researcher whose book *The Mothman Prophecies* retold the story of the Point Pleasant, West Virginia, encounters in which, for over a year, dire and bizarre prophecies and events such as UFOs, Men in Black, flying creatures, mutilated animals, and communications by a strange winged being all came true. The prophecies came to an end with the disastrous collapse of a bridge over the Ohio River. Keel's book was the basis of a feature film of the same name.

Keleti, George

Hungarian Minister of Defense who said, in Budapest on August 18, 1994, "Around Szolnok, many UFO reports have been received from the Ministry of Defense, which obviously and logically means that they (UFOs) know very well where they have to land and what they have to do. It is remarkable indeed that the Hungarian newspapers, and in general, newspapers everywhere, reject the reports of the authorities."

Kennedy, John F.

The thirty-fifth president of the United States, whose assassination resides at the pinnacle of conspiracy theories alongside that of the government's cover-up of UFOs. Not only did President Kennedy admit he knew something about UFOs that he could not disclose, but his alleged lover, Marilyn Monroe, also supposedly threatened to disclose to the press that the president had told her about the Roswell crash.

Kennedy also reported a 1963 UFO sighting off Cape Cod, near the family compound at Hyannisport, Massachusetts. The object was reportedly sixty feet in diameter and saucer-shaped, with a gray top and bright, silvery bottom. As it hovered over the water, it allegedly emitted a humming sound. There were plenty of witnesses along with the President, but JFK made those who saw it swear to keep silent.

Author Jim Marrs reported that he interviewed a steward aboard Air Force One named William Holden, who told him that he once asked the president what he thought about UFOs, but Kennedy said he couldn't say anything because his "hands were tied."

Although this story has been disputed, the late Philip Corso revealed that he had briefed Robert Kennedy on his work with the Roswell debris while at Army Research and Development, and that Robert Kennedy had briefed his brother.

One of the unsubstantiated stories regarding President Kennedy and UFOs concerns a speech he was scheduled to deliver shortly before his assassination, in which he was reportedly set to reveal the truth about Roswell, and that the aliens onboard were not enemies, but friends who promised a new world filled with prosperity and happiness for all. It was the potential release of this information, some have suggested, that triggered the CIA to order Kennedy's assassination.

Kennedy, Robert F.

New York senator and former U.S. Attorney General. One of the most intriguing stories about Robert Kennedy comes from the late Philip Corso, regarding secret Senate testimony he gave describing the CIA's protection of narcotics growers in Southeast Asia in the 1950s, and the spy agency's establishment of a secret "shadow government" that opposed the policies of President Kennedy. After naming names to a Senate subcommittee chaired by Strom Thurmond, Corso was contacted by Attorney General Kennedy, who wanted to know if Corso's charges were true. Corso not only convinced RFK that it was so, but that Corso was also engaged in top-secret research efforts to reverse-engineer technology from a flying saucer that had crashed in Roswell.

According to Corso, RFK took this information to his brother in early 1963. JFK was outraged at what the CIA paramilitary was doing and began formulating plans for withdrawing American military advisors from Vietnam and also swore to redouble his efforts to investigate UFOs. It was the president's "war" against the CIA, Corso stated in a unpublished manuscript, that lay behind not only JFK's assassination, but Robert Kennedy's assassination in 1968 as well.

Keyhoe, Donald

Marine Maj. Donald Keyhoe was the director of the National Investigations Committee on Aerial Phenomena (NICAP) and a strong believer in the existence of extraterrestrial UFOs.

Keyhoe, a fiction and aviation writer in later years, was a skeptic about flying saucers until 1949, when he was asked to write an article about them for *True* magazine after the June 1947 sightings across the United States made headlines. After interviewing sources in the Pentagon and learning that professional aviation observers had seen flying saucers at close range and reported on them, Keyhoe became a believer, and his 1950 article in *True* became widely popular and helped launched the American cultural fascination with flying

Annapolis midshipman Donald Keyhoe.

saucers. He later wrote the book *The Flying Saucers Are Real,* which became the first in a series of books he wrote about the subject.

In 1953 Keyhoe wrote the best-selling *Flying Saucers From Outer Space,* in which, citing Air Force intelligence reports given to him by his military contacts, he made a powerful case in support of his argument. A press release by his publisher featured the revelation that the Air Force had secret movie films of UFOs. A few years later, Keyhoe would take over the helm of NICAP and spearhead the drive for Congressional hearings on government secrecy about UFOs. But the government's involvement with the phenomenon was never fully revealed, Keyhoe believed, because the CIA had covered it up.

Keyhoe's attempts to tell the American people what he had discovered about UFOs and the government's active role in keeping their existence a secret were oftentimes censored. During one appearance on television in 1958, while he tried to explain that an open Congressional investigation would reveal what the government was hiding about flying saucers, the show was taken off the air for what were said to be "security reasons."

Keyhoe's long-running feud with the CIA over UFO disclosure probably accounted for his being removed as director of NICAP in 1969. *See* **Central Intelligence Agency, History of UFOs in; NICAP**

Keyhole Reconnaissance Satellites

A series of digital-imaging reconnaissance satellites that forms part of the U.S. defense satellite network, which continually picks up "fastwalkers," the codename for UFOs. *See* **Fastwalkers**

KGB Files

The much-feared and relentlessly efficient Soviet intelligence and counterintelligence service that came into being after World War II. Deep within the KGB files, UFO researchers believe, lie not only countless references to UFO intrusions over Soviet air space—intrusions that American radar and satellite photos confirm took place—but official Soviet responses to those intrusions. With the sale of the KGB files to Yale University, Soviet scientists and former cosmonauts were prompted to come forward to acknowledge the Soviet military's investigations into the nature of the UFO phenomenon. Among the many startling revelations stemming from the release of these files is the admission that both Russian and American intelligence services have assembled enough evidence to conclude that "there is a detachment of observers from other worlds traveling in near-earth orbit."

Khrunov, Yevegni

A Soyuz 5 pilot who was quoted in the article "UFOs Through the Eyes of Cosmonauts" as saying, "As regards UFOs, their presence cannot be denied: thousands of people have seen them. It may be that their source is optical effects, but some of their properties, for instance, their ability to change course by ninety degrees at great speed, simply stagger the imagination."

Kilgallen, Dorothy

Renowned national newspaper columnist and television personality, Kilgallen not only held one of the last interviews with Jack Ruby, the man who killed accused JFK assassin Lee Harvey Oswald, but she was also about to reveal secrets about the U.S. government's cover-up of UFOs. Apparently, Kilgallen had learned through her sources that the American government believed UFOs were real. This was deemed threatening enough to the CIA that it wiretapped her phone and placed her under surveillance.

In a May 23, 1955 article in the *Fort Worth Star-Telegram*, Kilgallen reported that the British had examined the debris from the wreckage of a flying saucer and become convinced that the objects "originated from another planet ... (and were) staffed by small men, probably under four feet tall." The article concluded by reporting the British believed "a flying ship of this type could not possibly have been constructed on earth." Kilgallen died under mysterious circumstances in 1965.

Kimberly, Earl of

In the UK House of Lords on January 18, 1979, the Earl of Kimberly said, "UFOs defy worldly logic. The human mind cannot comprehend UFO characteristics: their propulsion, their sudden appearance, their great speeds, their silence, their maneuvers, their apparent antigravity, their changing shapes."

Kingman, Arizona

An alleged May, 1953, alien landing wherein the aliens, rather than have their craft shot down by a radar pulse, simply left their vehicle intact for the military, ostensibly to give the military what it so actively sought: a working spacecraft. Eyewitness Bill Uhouse stated that a disk was left at Kingman with the hatch open. The first soldiers to enter it became disoriented and nauseous. The disk was transported to Area 51. "It made loud humming noises the whole time. The military didn't know what to do. They weren't sure if it was going to explode or what, so they left it sitting out in the open, on the tarmac for nine months." Uhouse was among many other people who have come forward since the 1950s to say they had worked alongside gray aliens at U.S. military bases for years. *See* **Area 51; Oz Effect; Uhouse, Bill**

Klass, Philip

A noted UFO debunker, author, and former aviation journalist who has spent many years researching UFOs and disputing the veracity of witness claims. According to a story by UFO researcher Ted Oliphant, he once asked Klass the question, "What do you think about the existence of UFOs?" Klass answered, "Oh, I suppose another civilization sends a probe here about every hundred years."

Famous debunker and UFO gadfly Philip Klass.

Knapp, George

A Las Vegas-based investigative reporter and commentator employed at KLAS-TV (CBS) in Las Vegas. Knapp has reported on such diverse topics as organized crime, government scandals, national politics, and UFOs. He was the first reporter to unveil the Area 51/UFO connection, and first publicized Bob Lazar's claims in his series of news reports, "UFOs: The Best Evidence" on KLAS-TV in November of 1989. After the fall of the Soviet Union in 1991, Knapp obtained hundreds of UFO-related documents from the Russian Ministry of Defense detailing the extensive UFO studies conducted by the Russian government, military, and scientific establishment, including the "Russian Roswell" case. *See* **Lazar, Robert; Russian Roswell Case**

Kovalyonok, Vladimir

A Soviet cosmonaut who was quoted in an interview with an Italian news correspondent as saying:

On May 5, 1981 we were in orbit (in the Salyut-6 space station). I saw an object that didn't resemble any cosmic objects I'm familiar with. It was a round object which resembled a melon, round and a little bit elongated. In front of this object was something that resembled a gyrating depressed cone. I can draw it. It's difficult to describe. The object resembles a barbell. I saw it becoming transparent and like with a body inside. At the other end I saw something like gas discharging, like a reactive object.

Then something happened that is very difficult for me to describe from the point of view of physics. Last year in the magazine *Nature* I read about a physicist—we tried together to explain this phenomenon and we decided it was a plasma form. I have to recognize that it did not have an artificial origin. I don't know of anything that can make this movement—tightening, then expanding, pulsating. Then as I was observing, something happened, two explosions. One explosion, and then 0.5 seconds later, the second part exploded. I called my colleague Viktor (Savinykh), but he didn't arrive in time to see anything.

What are the particulars? First conclusion: The object moved in a sub-orbital path, otherwise I wouldn't have been able to see it. There were two clouds, like smoke, that formed a barbell. It came near me and I watched it. Then we entered into the shade for two or three minutes after this happened. When we came out of the shade we didn't see anything. But during a certain time, we and the craft were moving together.

See **Salyut 6 Space Station**

Krapf, Philip H.

Philip Krapf

A former editor at the *Los Angeles Times* who has claimed to be a contactee visited by extraterrestrials who've disclosed secrets of the universe to him. Krapf said he was beamed aboard an extraterrestrial spaceship for an encounter with alien beings that lasted several days. He was told there was a worldwide plan going on to prepare the people of earth for the inevitable contact the human race would have with these and other beings in the universe. Krapf was also told he was to play a key role in this process, along with thousands of other hand-picked people around the globe. He was picked, the aliens told him, because of his complete lack of interest and knowledge about UFOs, as well as his skill as a writer and editor.

Krapf reveals in his book *The Contact Has Begun* that key people in the media already know the truth about flying saucers and have agreed not to release the story.

KSWS, Roswell Radio Station

An ABC and Mutual Broadcasting affiliate. Station manager John McBoyle visited the Roswell crash site and called his teletype operator, Lydia Sleppy, to tell her to put the story out over the wire. However, before she completed the transmission, she says, she received a teletype from the FBI telling her, "This is the FBI, you will cease transmitting." *See* **McBoyle, John; Roswell, New Mexico; Sleppy, Lydia**

Kubitschek, Juscelino

The former president of Brazil who confirmed in 1958 that the Brazilian Navy had indeed seen and photographed several flying saucers fly by their ships during a Naval exercise. Kubitschek said that the navy had analyzed the photos, found them to be authentic, and that he, too, vouched for their authenticity. *See* Brazilian Navy Case

Kuzovkin, Aleksandr

In an article in the June 1989 issue of *Soviet Military Review*, Kuzovkin, commenting on the threat of a UFO formation being misidentified as a ballistic missile launch and triggering a counter-strike by either side, wrote:

> We believe that lack of information on the characteristics and influence of UFOs increase the threat of incorrect identification. Then mass transition of UFOs along trajectories close to that of combat missiles could be regarded by computers as an attack.

Laser Weapons

Lasers (Light Amplification by Stimulated Emission of Radiation) have been incorporated into some highly regarded anti-ballistic missile defense systems foreseen as crucial to future warfare. Although the U.S. military won't acknowledge it, beam technology—both in the form of lasers and directed particle-beam weapons—has been acknowledged as the most commonly reported weapons system derived from the reverse-engineering of UFO craft. Of the twenty thousand precision-guided munitions used in the Gulf War, over 60 percent were laser-guided. Among the weapons unleashed upon Iraq in Gulf War II, almost all of them relied on technologies that ultimately derived from the flying saucer crash at Roswell.

In 1997, the MIRACL (Mid-Infrared Advanced Chemical Laser) was tested by the Pentagon against a soon-to-be retired American scientific satellite. The test succeeded in blinding the satellite, rendering it useless. This heralded the deployment of lasers as anti-satellite as well as anti-missile weapons, enabling the U.S. military to secure near-earth-orbit supremacy by being able to destroy enemy satellites and at the same time protecting its own from anti-satellite killers.

The next stage is the deployment of an airborne laser mounted on a high-flying militarized Boeing 747 jetliner, followed by a higher-powered version carried aloft on an advanced (and still classified) flying triangle neutral-buoyancy craft. These lasers will be able to target and destroy just-launched ballistic missiles or even cruise missiles flying at lower altitudes. The laser's value comes from its almost limitless range and high power, and its ability to acquire targets, point and shoot, and reacquire in mere seconds. Married to battlefield management systems like the Navy's AEGIS, they are a formidable weapon.

Lawrence, C.D.

In April 1897 Lawrence photographed a large cigar-shaped object floating over Chicago. This was seven years before the first dirigible flew in the United States and six years before the famous flight of the Wright Brothers at Kitty Hawk. *See* **Great Airship Mystery**

Lazar, Robert

One of the more enigmatic figures in current UFO research and lore, Robert Lazar has claimed that he worked at site S-4 at the mysterious and top-secret Area 51 in Nevada, on a project that involved the reverse-engineering of extraterrestrial technology.

Robert Lazar, who claimed to have worked at Area 51 on alien propulsion systems.

Lazar, who has been described as an expert in propulsion systems, alleged that he had been hired from the Los Alamos Meson Physics Facility to work at S-4 as part of a team that was attempting to re-engineer propulsion technology from a captured alien saucer in order to use it in a man-made craft. During his tour of duty at S-4, Lazar said that he had been ordered to evaluate different types of flying saucers so he could determine how to re-create the crafts' propulsion systems; that he witnessed one of the disks on a test flight over the desert; and that he observed what he thought was an extraterrestrial working among human scientists. Of necessity, Lazar's work was so highly classified that he was continually subjected to surveillance and background checks, and he would often run afoul of S-4 officials for "minor" violations of protocol and security.

Lazar subsequently invited his wife and friends out to the test range, on public land, to observe a flying saucer test flight and was eventually reprimanded and threatened with prosecution. After that, he went public, telling his story to reporter George Knapp of television station KLAS in Las Vegas. *See* **Area 51**

Lear, John

John Lear

Son of Lear Jet founder William Lear. A pilot and contract employee of the CIA, flying for the Agency during the Vietnam War, Lear has won speed records in the Lear Jet.

Lear gained notoriety by circulating a story, repeated since by others, that the current Majestic-12 Group was still managing our government's response to an extraterrestrial presence and was in charge of a section at Area 51 where research on aliens was taking place. Lear has said that the U.S. government entered into a deal with the extraterrestrials under the terms of which for a conveyance to the US government of advanced ET technology, our government would allow the ETs to conduct experiments on human beings they abducted and conduct biological experiments on livestock they captured.

Although Lear's claims have been criticized by others both within and outside the UFO community, other people, including some government whistleblowers, have cited Lear as having told the truth about the larger issues of secret government experiments on alien technology taking place at Area 51. *See* **Area 52; Majestic**

Leir, Roger

This Southern California doctor has become known as the "Alien Implant Surgeon." His extractions of metallic devices from the bodies of individuals who claim to be abductees has caused much controversy within the UFO community.

Dr. Leir, a board-certified podiatric surgeon, was originally skeptical about the reports of abductees who claimed to have had medical experiments performed on them and objects inserted under their skins by their alien examiners. But after reviewing the medical records and X-rays of an abductee who asked Leir to remove just such an object and performing surgery on the patient, Leir became convinced that these devices were real.

Leir discovered that the object was not only implanted well beneath the patient's skin without any obvious point of entry, but there was no damage to the surrounding tissue and no rejection of the object by the body's natural defenses. A membrane surrounding the object itself was found to be composed primarily of the patient's own neural tissue connected directly to the patient's nervous system—something Leir had never seen before.

Roger Leir

When he touched the object with a probe, the heavily anesthetized patient awoke and almost jumped off the operating table. Subsequent tests performed on this object showed it to be composed of elements in proportions not naturally found on earth. One testing lab suggested to Dr. Leir that the object had originated in outer space, possibly from a meteorite.

Dr. Leir's experiments and investigations into UFO encounters continue, and his book and video, *The Alien and the Scalpel,* documenting the surgeries he performs, have continued to arouse both interest and controversy within the UFO community. *See* **Implants, Alien**

Leonard, George

A researcher and writer who, in his 1977 book *Somebody Else Is on the Moon*, claimed that there are artificial structures on the moon, placed there by extraterrestrials. Leonard has said, "The moon is occupied by an intelligent race or races which probably moved in from outside the solar system. The moon is firmly in possession of these occupants." *See* **Moon, Dark Side of the**

Leslie, Melinda

A researcher and victim of military abductions, Leslie has also experimented with out-of-body projection/remote viewing. Leslie's special area of interest has been to delve into the mysteries that lie behind the motivations of military operations that conduct abductions of private citizens, who are then subjected to various programming procedures. She has suggested that the military is working with alien creatures to improve their

Melinda Leslie

mind-probe and programming techniques. Leslie believes that abductees are also pawns between their ET and military abductors, as each side tries to find out what the other has learned. *See* **Abductions, Indicators of; Military Abductions (MILABS)**

Lockheed

An aerospace defense contractor whose "Skunk Works" division developed top-secret high-performance aircraft such as the U-2, SR-71, and F-117 Stealth fighter—and aircraft that the public still doesn't know about. The former head of Lockheed, Ben Rich, once said that the U.S. possessed a level of technology well beyond what the public believed existed, and that it not only knew about ETs, but had the capability to send them back to their home planet. *See* **Johnson, Kelly; Rich, Ben; Skunk Works**

Loman, Ted

A UFO researcher who has attempted to confirm the dates of alleged flying saucer crashes by locating specific issues of newspapers in different local libraries and cataloging the reports contained therein. Loman discovered that in many cases specific back issues on the dates of the alleged crashes were missing even though issues before and after the key date were on file.

He has said it was as if someone had systematically visited each and every newspaper morgue that might hold a record of a UFO crash and removed the newspaper issues, thereby destroying all reports of the incident.

Lonehood, Lex

Webmaster and web producer of coasttocoastam.com, the website affiliated with the UFO and paranormal-themed overnight radio talk show, "Coast to Coast AM," made popular by Art Bell and George Noory. An artist and writer, Lonehood contributes feature-length articles and columns to *After Dark,* the monthly magazine associated with the show.

Lex Lonehood, webmaster for the radio program Coast to Coast AM.

Lonehood create the webzine, "oFFbEAt" for the online division of the *San Francisco Chronicle* in 1995. *See* **Bell, Art; Noory, George**

Los Alamos National Laboratory

A highly classified nuclear and scientific research facility in New Mexico, where, during World War II, the world's first atomic weapon was developed. The Los Alamos site might have been one of the attractions for the flying saucers that flew over Roswell in 1947.

Los Alamos is also where Robert Lazar claimed to have worked before getting a job at Area 51, and its metallurgy lab was where Dr. Roger Leir brought his samples of extracted alien-implant devices for study, resulting in the finding that they contained isotopes of elements to a degree not usually found on earth. *See* **Lazar, Robert; Leir, Roger**

Los Angeles Air Raid of 1942

On February 24, 1942, a formation of "unknown" aircraft crossed the California coast at Long Beach and were spotted by hundreds of witnesses—some of whom were professional and credible observers—and fired on by anti-aircraft batteries. The craft, allegedly traveling at almost five miles a second, avoided the bursting shells and proceeded east and north towards Culver City. There was also a report of a large mothership-type of craft, moving slowly, which took a number of direct hits from anti-aircraft barrages and apparently suffered no damage, thus disproving claims that it could have been a surveillance dirigible or a Japanese balloon-bomb. *See* **Collins, Paul**

Lovell, James

Gemini 7 and Apollo 8 astronaut whose radio transmission of a "bogey" tracking him during the Gemini 7 mission and a "Santa Claus" shadowing him on the Apollo mission are historic pieces of evidence that the astronauts sometimes routinely encountered UFOs on their space flights and lunar landings. As the astronauts emerged from the dark side of the moon, Lovell radioed: "Please be advised that there is a Santa Claus." *Santa Claus* was code for a UFO. *See* **Apollo Program; Borman, Frank; Schirra, Wally**

Low, Robert

Project coordinator for the 1969 Condon Report on UFOs who wrote a memo suggesting that the evaluators of the submitted evidence be predisposed to disbelieve what they were about to see and hear. This is said to have tainted the report from the very beginning. Low wrote that "the trick" would be to convince the public and anyone else reviewing the material that the study was completely objective, although it was actually rigged so that there would be "zero expectation of finding a saucer." The Condon Report was a set-up by the U.S. Air Force to provide a pretext for closing down Project Blue Book, which they did in 1969. *See* **Condon Report; Projects, Government UFO**

LSD, Experiments

Conducted upon unsuspecting Army personnel, these experiments with the hallucinogenic chemical LSD were part of an official government study into non-lethal weapons that would disorient large numbers of an enemy population as well as their military leaders so as to make it easier for the U.S. to defeat enemy armies in war and occupy hostile territories.

LSD experimentation was part of a long-term military and CIA project in mind control as well, which sought ways to alter the belief systems of key targets to make them susceptible to suggestions from their handlers. See **Mind Control; Military Abductions (MILABS)**

Lubbock Lights

In 1951, over forty-five years before the Phoenix Lights floated over that desert community, the town of Lubbock, Texas was visited by a similar formation of strange objects lighting up the night sky. Although scientists at the time claimed that the UFOs were, in fact, flocks of migrating birds whose bellies reflected the nearby street lights, photographs taken of the objects seemed to indicate the objects were aircraft of some kind, not birds.

Like the Phoenix Lights, the Lubbock Lights became the subject of debate between non-believers, determined to deny all evidence of anything paranormal, and true believers, convinced that they had tangible scientific evidence of a UFO sighting. *See* **Phoenix Lights**

Luna 9

A Soviet moon-landing probe of the 1960s which, while still in lunar orbit, photographed what looked to be long runway formations, pyramid-shaped objects, and other apparent artificial structures. Luna 9's gravitational readings also suggested that the moon might have a hollow core, with several large centers of localized intense gravity just below the surface as if there were huge structures built underground. *See* **Moon, Dark Side of the**

"Lunar Lights"

Anomalous events documented in NASA's study of five hundred years of lunar observations, manifested usually as lights moving across lunar craters at speeds up to six thousand miles per hour or hovering over localized areas of the moon for days or weeks at a time before disappearing. These events have been observed by sky watchers during the Elizabethan period in England and during the early colonial period in the U.S.

They have been recorded by the British Royal Observatory in the nineteenth century and were noted by German scientists in the early twentieth century. In addition to the lights, observers have also recorded what appear to be clouds of fog and mist, elongated beams of light, and "rays" all emanating from the moon's surface. *See* **Moon, Dark Side of the; National Aeronautics and Space Administration Technical Report # 277**

Lunar Orbiter II

Launched in 1966, this American lunar probe reportedly took many photos of the ruins of an artificial city on the moon, including obelisks and other large geometric structures. These photographs of the Sea of Tranquillity have remained classified.

Lunar Orbiter probe.

Lunar Orbiter III

According to some UFO researchers, Lunar Orbiter III took photos of large runway formations on the moon as well as of the "Shard." The Shard's existence is inexplicable as a naturally occurring lunar feature. *See* **Moon, Dark Side of the; Shard, The**

Lure, Project

It has been suggested that this project, mentioned by Donald Keyhoe in his book *Aliens from Space*, was the basis for the motion picture *Close Encounters of the Third Kind,* the story of a scientific project to encourage a flying saucer to land at a specific spot. Actually, Lure was a Canadian project to build a UFO landing site. The site was built, but no UFOs ever landed there.

MacArthur, Douglas

Supreme commander of Allied forces in the Pacific during World War II and of the United Nations forces in the Korean War. Gen. MacArthur was quoted in the *New York Times* of October 8, 1955 as saying, "Because of the developments of science, all the countries on earth will have to unite to survive and make a common front against attack by people from other planets. The politics of the future will be cosmic or interplanetary." He was also quoted as saying that the next world war will be fought in outer space.

Maccabee, Bruce

UFO researcher and *UFO Magazine* feature writer Bruce Maccabee received his Ph.D. in physics from American University and has worked at the Naval Surface Warfare Center since 1972. Active in UFO research since the 1960s when he joined NICAP, he has since became a member of MUFON and was involved in the investigation of the Gulf Breeze Sightings and the claims of Gulf Breeze contactees Ed and Frances Walters.

He is currently the State Director of Maryland MUFON. Dr. Maccabee's research also figured prominently in the investigation of the Phoenix Lights incidents in 1997. He has written *The UFO/FBI Connection* about the real FBI X-Files, having to do with their investigation of the UFO phenomenon as far back as 1952, and *Abduction In My Life*.

Mack, John

UFO researcher, professor of psychiatry at Cambridge Hospital, Harvard Medical School, a practicing physician, and the founding director of the Center for Psychology and Social Change, Dr. Mack is also the Pulitzer Prize–winning author of *A Prince of Our Disorder*. His work with patients who claim they've been abducted by aliens has convinced him that they fervently believe in the truth of their experiences.

Mack's findings have caused rumblings within the psychiatric community and amongst his more conservative peers, as well as the Harvard University board of directors. He has weathered the storm, however, and continues his controversial practice.

In his book *Abduction: Human Encounters With Aliens*, Mack chronicles thirteen of his most interesting cases. From them, he has been able to classify distinct types of information with respect to the phenomenon, regarding as important the physical evidence left behind such as burned grass, scars, lesions, bruises, or implants—things that science can evaluate. He has also seen it important that humanity continues its attempts at communicating with unknown life forms who live on levels of consciousness where communication goes beyond spoken language. He believes the investigation of human abductions isn't just about aliens or UFOs; it's really about how humanity can expand its understanding of reality and awaken deeper human potential. *See* **Abduction Scenario, Physical Evidence**

MacKenzie, Steve

A witness to the Roswell crash-debris removal, MacKenzie signed a sworn affidavit to the fact. On July 2, 1947, he was ordered to report to White Sands, New Mexico, where he radar-monitored fast-moving UFOs flying over the area. On July 5, while back at Roswell, he was told to accompany a special team from Washington to the crash site. MacKenzie said that upon arrival they realized it was one of the objects they had tracked on radar: "We were all smoking cigarettes and talking about how in the hell we were going to handle this thing. We were all concerned and a little scared." MacKenzie also confirmed that the Army completely "sanitized" the site, even using vacuum cleaners to complete the job. He claims to have personally observed four small bodies at the crash scene, one still inside the craft and sprawled across a seat. MacKenzie later learned there was a fifth body at the site.

Magdalena, New Mexico

The town mentioned in the Ramey telegram photographed by J. Bond Johnson as a site of an alien spacecraft crash. *See* **Ramey Telegram; Roswell, New Mexico**

Majestic

The codename of the alleged top-secret project, established by President Harry Truman, to investigate and deal with the Roswell incident. MJ-12 was also the name of the group tasked to study the importance of the UFO phenomenon in general, to cover up its reality from the public, to determine how to handle the implications of this find, and to prepare incoming president Dwight Eisenhower for dealing with the stunning reality that an extraterrestrial spacecraft had crashed on earth.

The first set of MJ-12 documents, which turned up in the mail of Hollywood producer Jaime Shandera, describe the fact that the

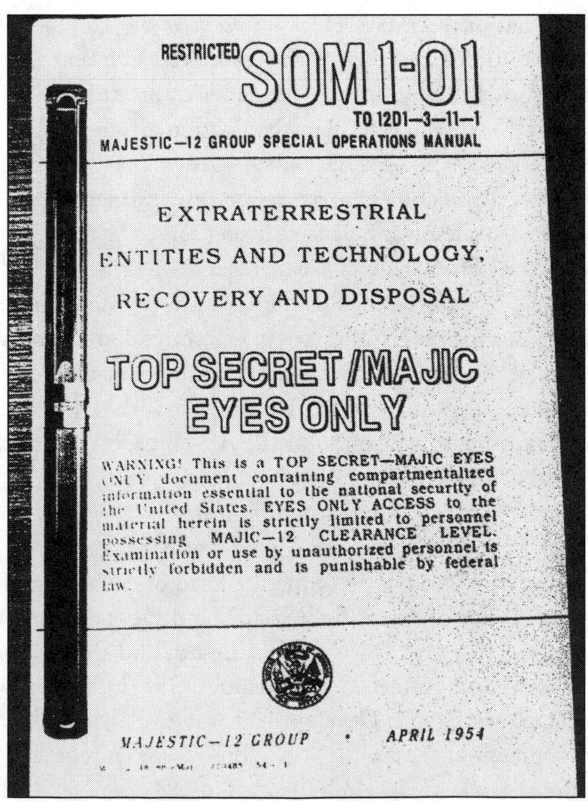

RESTRICTED SOM 1-01

TO 12D1—3—11—1

MAJESTIC—12 GROUP SPECIAL OPERATIONS MANUAL

EXTRATERRESTRIAL
ENTITIES AND TECHNOLOGY,
RECOVERY AND DISPOSAL

TOP SECRET/MAJIC
EYES ONLY

WARNING! This is a TOP SECRET—MAJIC EYES ONLY document containing compartmentalized information essential to the national security of the United States. EYES ONLY ACCESS to the material herein is strictly limited to personnel possessing MAJIC—12 CLEARANCE LEVEL. Examination or use by unauthorized personnel is strictly forbidden and is punishable by federal law.

MAJESTIC—12 GROUP • APRIL 1954

Alleged MJ-12 manual for dealing with Extraterrestrials

U.S. military recovered a crashed alien spacecraft and bodies outside of Roswell, New Mexico, in 1947. They also state that another crash occurred in Texas near the Mexican border in 1950, and that alien bodies were recovered from that wreckage as well. The documentation also includes a list of MJ-12's participating members.

Although the authenticity of the documents and the existence of a group called Majestic has been the subject of fierce debate within the UFO community, author Stanton Friedman has argued convincingly that, even on circumstantial evidence alone, the existence of such a group must be true in order for the subject of UFOs to have been studied, dealt with, and then covered up. According to Friedman, the members of the group were:

Dr. Lloyd V. Berkener

Explorer and scientist. Berkener was a member of a CIA panel that determined that UFOs did not constitute a threat to U.S. national security. He was executive director at Carnegie Institute, a member of the Joint Research and Development Board in 1946, and headed a committee that later became the Weapons Systems Evaluation Group.

Dr. Detlev Bronk

Physiologist, aviation expert, and prominent member of the National Academy of Sciences, Bronk served as president of both Johns Hopkins and Rockefeller Universities. He was chairman of the Nuclear Research Committee and the medical advisor for the Atomic Energy Commission. Bronk also served on the Scientific Advisory Committee of Brookhaven National Laboratory where he worked with Dr. Edward Condon, who, at the behest of the Air Force, became director of the Condon Committee.

Dr. Vannevar Bush

A leader in research and development at MIT and Carnegie Institute, Bush was at one time the head of each of the following: Office of Scientific Research and Development, the Joint Research and Development Board, and the National Advisory Committee on Aeronautics. The Office of Scientific Research and Development was responsible for the development of the atomic bomb. Bush also put together the National Defense Research Council in 1941.

James Forrestal

Former Princeton graduate and Secretary of the Navy, Forrestal later became the first Secretary of Defense in 1947 during the time of the Roswell crash. His apparent suicide in 1949 came shortly after his commitment to Bethesda Naval Hospital for reported "depression." Forrestal's suicide has been questioned by some researchers who believe he wanted to "go public" with the truth about flying saucers, and by others who think the alien "reality" and its threat to mankind were too much for him to take.

Gordon Gray

Secretary of the Army, Gray later held several high security positions under Presidents Truman and Eisenhower. He was also a consultant on UFOs and reported directly to CIA Director Walter B. Smith.

Admiral Roscoe Hillenkoetter

First director of the Central Intelligence Agency (CIA). Hillenkoetter later became a member of NICAP, the National Investigations Committee on Aerial Phenomena. He publicly stated that "UFOs are real ... [though] through official secrecy and ridicule, many citizens are led to believe the unknown flying objects are nonsense."

Dr. Jerome Hunsaker

Head of the National Advisory Committee on Aeronautics and chairman of the departments of mechanical and aeronautical engineering at MIT.

Dr. Donald Menzel

Astronomer, director of the Harvard College Observatory, and an expert in cryptoanalysis, the science of breaking codes and deciphering unknown languages and symbols. Menzel was involved in high-level intelligence operations during and after World War II. As Stanton Friedman explains in his book *Majec/Top Secret*, Menzel's publicly stated views that UFOs were a fiction was also a cover for his role on MJ-12.

General Robert M. Montague

Head of the Armed Forces Special Weapons Center, Montague was also an Army general at Fort Bliss and, as such, had control over the White Sands nuclear research facility and the Sandia Atomic Energy Commission facility at Albuquerque, New Mexico, during the time of the saucer crash at Roswell.

Admiral Sidney Souers

The first director of Central Intelligence, the precursor to the CIA. Souers was also a member of the National Security Council and a special consultant for military intelligence operations.

General Nathan Twining

Chairman of the Joint Chiefs of Staff, Chief of Staff of the U.S. Air Force (1953–1957). Twining was also the commander of the Air Materiel Command based at Wright-Patterson, where the debris and bodies from the Roswell crash were taken. Twining made a sudden appearance at Roswell on July 8, 1947, after canceling a scheduled trip, and the same day Roswell Army Air Base sent out a press release stating that a flying disk had been recovered. In a now- famous memo, Twining stated that UFOs were very real.

General Hoyt Vandenberg

Chief of Staff of the U.S. Air Force (1948–1953) and former director of Central Intelligence, he was allegedly in charge of security for the MJ-12 group.

Maltais, Jean and Vern

Close friends of Grady Barnett to whom he told his story about the San Augustin crash site and the alien creatures and spacecraft he saw there. *See* **Barnett, Grady**

Maltsev, Igor

Former chief of the Soviet Air Defense Forces who has been frequently quoted about his belief that UFOs exist and that governments of the world know of their presence. In a U.S. Department of Defense memo, the writer cites Gen. Maltsev's revelations about UFOs in the USSR and confirms his belief that intelligent beings from outer space have been visiting the Soviet Union. *See* ***Rabochaya Tribuna;*** **Russian Roswell**

Mansfield, Ohio, Army Helicopter Case

On October 18, 1973, a UH-1 Army Reserve helicopter and its four-man crew (Capt. Lawrence Coyne, Lt. Arrigo Jezzi, Sgt. John Healey, and Sgt. Robert Yanacsek) were returning at night to their home base of Cleveland from Columbus, when they encountered something strange in the air ten miles southeast of Mansfield, Ohio. At 2,500 feet altitude, Sgt. Yanacsek observed a distant red light. Thirty seconds later the light moved rapidly towards the helicopter, forcing Capt. Coyne to put his aircraft into a steep dive to avoid a collision. At 1,700 feet altitude, just as the Army crew braced themselves for impact, the object abruptly stopped in mid-air and began to keep pace with the helicopter at close range. The crew was then able to observe the object: a sixty-foot-long, cigar-shaped craft with a low dome on top. It had no tail or wings, made no sound, and created no turbulence from its close proximity. Yanacsek later said he thought he saw a hint of dimly-lit windows between the main body and the dome, and Healey described the body of the object as "gun-metal gray."

The object then rose above the UH-1 and shot out a green beam of light, enveloping the helicopter and pulling it upwards, despite the crew's attempts to fight the controls. Suddenly, the UH-1 was released, and the crew brought her back under control and were finally able to land in Cleveland.

Although skeptic Philip Klass tried to pick this story apart, other researchers have successfully refuted Klass' debunking, and the helicopter flight crew also received corroboration of their account from ground observers. The Mansfield helicopter controversy was presented to the Sturrock Rockefeller Panel that convened at Pocantico Hills, New York, in July 1998. The panel deemed the story intriguing enough to warrant further investigation. *See* **Sturrock Report; Rockefeller Panel**

Mantell, Thomas F.

The January 9, 1948, *New York Times* story read, "Flier dies chasing a flying saucer." The flier was Thomas F. Mantell, a Kentucky Air National Guard captain flying an F-51 fighter alongside two other Air National Guard planes. All three pilots had been scrambled to intercept a large UFO moving slowly overhead, a response triggered by the sighting reports of the Kentucky State Police and many citizens on the ground.

Mantell radioed ground control, "I'm closing in for a better look," as he and the other pilots climbed to intercept the UFO. None of the planes were equipped with oxygen, and they were running low on fuel. The other pilots turned back after reaching 22,000 feet in accordance with Air National Guard regulations that required pilots to use oxygen above 14,000 feet. Mantell, however, continued to climb to 30,000 feet. At some time during the pursuit, Mantell's plane went into a spin and plunged to the ground. Mantell was killed. He was still strapped to his seat and his watch had stopped at 3:18 P.M., the time of impact.

Initially it appeared that Mantell had simply chased a weather balloon too high, lost consciousness from a lack of oxygen, and crashed. However, over the intervening years, UFO researchers have noted some strange circumstances surrounding the pilot's death:

- Mantell's military ground controllers knew how to tell the difference between a weather balloon and another object. Had Mantell been chasing a balloon, he would have been ordered back to the field immediately after reaching 14,000 feet.
- Mantell's last radio transmission was reportedly, "My God, I see people in this thing!"
- According to Mantell's instrument panel, his aircraft had remained aloft for over an hour after running out of fuel.
- A farmer who witnessed the crash said that as the plane fell toward the ground, it was enveloped in a bright white light. It then fell out of the light a short distance from the ground, no longer in a tailspin. This would explain why the recovered aircraft was almost undamaged.
- Mantell died because his shoulder straps broke. He was thrown forward and the control stick punctured his chest. Had the straps not broken, judging from the remarkably undamaged condition of the plane, Mantell might well have survived. Mantell's death, the first recorded case in modern ufology, remains unsolved to this day.

Marcel, Jesse A.

The base intelligence officer of the 509th Bomber Squadron at Roswell Army Air Field. Maj. Marcel was ordered by base commander Col. William Blanchard to accompany rancher Mac Brazel back to the Foster Ranch to survey the wreckage of whatever crashed there. Marcel also described the material he picked up and brought back to his house to show his family. In an interview in 1979, he said:

When we arrived at the crash site, it was amazing to see the vast amount of area it covered; scattered over an area of about three-quarters of a mile long, I would say, and fairly wide, several hundred feet wide. So we proceeded to pick up all the fragments we could find and load up our Jeep Carry-All.

It was quite obvious to me, [my] familiarity with air activities, that it was not a weather balloon, nor was it an airplane or a missile. It was something I had never seen before, and I was pretty familiar with all air activities.

We loaded up the Carry-All, but I wasn't satisfied. I told Cavitt, "You drive this vehicle back to the base and I'll go back out there and pick up as much as I can and put in the car," which I did. But we only picked up a small portion of the material that was there.

A lot of it had a lot of little members with symbols that we had to call hieroglyphics because I could not interpret them, they could not be read, they were just symbols. Those symbols were pink and purple and lavender. I even tried to burn that. It would not burn. But something that is more astonishing is that the piece of metal that we brought back was so thin, just like tinfoil in a pack of cigarette paper. So I tried to bend the stuff, it wouldn't bend. We even tried making a dent in it with a sixteen-pound sledgehammer. And there was still no dent in it.

Jesse Marcel is at the center of the Roswell controversy because he was one of he first people to go public with the story of what he believed really happened. He told his story to Stanton Friedman in 1978, and that's when the disclosure process began. Marcel is a central, almost irrefutable figure because as the base intelligence officer, he would have been familiar with weather balloons as well as any other intelligence-gathering apparatus.

Whether or not he would have known what a Mogul balloon was is open to question, but he should have been able to identify weather balloon materials in any event. Yet, in his own words, nothing he recovered resembled anything he had ever seen before. As an account by one of the first eyewitnesses to the unadulterated Roswell debris, Marcel's testimony has withstood the test of time. *See* **Roswell, New Mexico**

Marcel, Jesse, Jr.

Major Jesse Marcel's son, who also saw the recovered material from the Roswell crash site when his father brought it home on the way back to his base. Jesse Marcel, Jr. became a doctor and the surgeon general for the state of Montana and remains a credible source within the UFO community as one of the few living individuals who, via independent verification, has seen the Roswell debris and can accurately describe it. *See* **Roswell, New Mexico**

Jesse Marcel, Jr., the son of Major Jesse Marcel, has said that he personally handled debris from the crash UFO outside of Roswell.

Marchetti, Victor

The former special assistant to the executive director of the CIA in 1979, who was quoted as saying:

> We have indeed been contacted, perhaps even visited, by extraterrestrial beings, and the U.S. government, in collusion with the other national powers of the earth, is determined to keep this information from the general public. The purpose of the international conspiracy is to maintain a workable stability among the nations of the world and for them, in turn, to retain institutional control over their respective populations. Thus, for these governments to admit there are beings from outer space—with mentalities and technological capabilities obviously far superior to ours—could, once fully perceived by the average person, erode the foundations of the earth's traditional power structure.

Marrs, Jim

Author, newspaper reporter, and researcher who first wrote about the assassination of President Kennedy in *Crossfire*, and about UFOs and alien abductions in *Alien Agenda*. *See* **Great Airship Mystery; Kennedy, John F.**

Mars

The real-life mysteries of the Red Planet could easily have been snatched from the pages of fictional lore, much as they remind us of the "Barsoom" of Edgar Rice Burroughs' *John Carter of Mars*, and Ray Bradbury's ethereal aliens of *The Martian Chronicles*. Earth's second-nearest planetary neighbor had once been thought to harbor intelligent life (astronomer Percival Lowell's "canal builders"), then found to be a cold, barren desert (Mariner 4, Vikings I and II), then "rediscovered" by more recent explorations (Mars Global Surveyor, Mars Odyssey, Mars Pathfinder, the Mars "rock") to possibly contain a rudimentary form of bacterial life, as well as liquid water.

But how does one explain the enigmatic "Face on Mars" and the Cydonian Pyramids, or the inexplicable loss of so many of our space probes over the decades—failures for which NASA and JPL scientists have only managed weak explanations? The lure of Mars has always been manifest in the similarities it shares with the earth and the obvious fact that it is, at the same time, so alien. How could humans not be fascinated with the Red Planet and *not* want to explore its two faces—which remain the stuff of legend, and of scientific fact?

In trying to understand the notion of the Face as artificial *vs.* the Face as natural, it is essential to compare the geological features on Mars to those on earth to ascertain whether natural forces could have produced the Face. On earth, geological features are highly dynamic, due to the forces of weather. In addition, the molten material inside the earth slowly shifts the continents. Plants affect landforms, slowly turning rock into dirt, which is then washed away by rains.

Martian geology is utterly different. First of all, Mars is very cold, and it has a very thin atmosphere. High temperatures on Mars are near the freezing point of water, and the temperature cools rapidly with distance from the ground. Mars does appear to have once had liquid water, which carved river and flood-like channels. Recent discoveries suggest that Mars may even have liquid water just

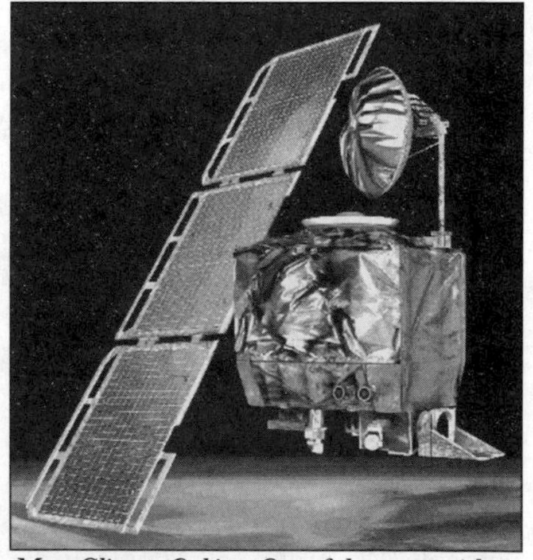

Mars Climate Orbiter. One of the many mishaps to befall Martian exploratory devices. This device failed to achieve Mars orbit, 1999.

below the surface. However, most scientists believe the water of Mars has been frozen into the soil. Occasionally, molten lava stirs beneath the surface and melts the frozen water, causing catastrophic floods, which leave behind chaotic terrain from the fallen blocks of the former surface. These blocks often look like jumbled pyramids.

Martian surface features are also affected by volcanism. Volcanoes on Mars are vast beyond earthly experience. Olympus Mons, a shield volcano like those found in Hawaii, is nearly the size of Texas, and Alba Patera, a type of volcano almost unique to Mars, is even larger. Huge lava flows cover vast areas of Mars, creating channels and ridges, but even more effective in creating strange land-forms is the persistent and endless wind. Areas of fretted terrain are pockmarked and scabbed by the effects of the wind and the small particles it carries, acting over millions and billions of years, in a way that would be impossible in earth's more dynamic environment.

Therefore, platforms and soft-sided features are not rare on Mars. There are many different kinds of mesas, and many features are etched by ice, water, lava, or wind over millions or billions of years. These are the sort of features which could easily become a "face."

As much as it would be intriguing to learn that the Face and the pyramids are the ruins of a former civilization on Mars or even a civilization that once existed on the surface but has now moved underground because of the loss of Martian atmosphere, science may have to content itself with the nevertheless dramatic discovery that life on Mars does exist in water beneath the surface or in tidal pools just above freezing on the surface, but that life is composed of small organisms that have managed to hang on.

One of the controversies surrounding the nature of Mars as a life-sustaining planet arose from the tests conducted by the twin Viking missions to Mars in 1976, when tests turned out to be suggestive that there was organic life in the Martian soil. Although the results of this test were later discounted by other evidence, it is not widely known that science could have made a case for life on Mars nearly a quarter-century ago. The test carried out during the Viking missions was based on a metabolic definition of life. In other words, if certain nutrients were fed into a soil sample, a chemical reaction in the soil might indicate the presence of organisms that had processed the nutrients. When the soil sample, in fact, yielded carbon dioxide as a result of the nutrient test, some scientists believed that constituted evidence of some sort of organic matter being present; that a form of metabolism had taken place.

Lending credence to the possibility of either past or present life on Mars was the June, 2003 discovery of an ice strata just beneath the surface of the soil on the Martian north pole. A team of U.S. and Russian scientists, analyzing data from Mars Global Surveyor, determined that the north polar ice cap on Mars has huge amounts of hydrogen, enough to indicate the presence of ice, just inches beneath the soil. This means that if there is evidence of life on Mars, from a time when the planet was wet and warm, it will probably turn up there. *See* **Mars, Face on**

Mars Climate Orbiter

The failure of the 1999 Mars Orbiter mission proved another bitter pill for NASA to swallow, coming as it did on the heels of the Mars Polar Lander debacle (1999) and the Mars Observer fiasco (1992). NASA and JPL cited the cause as a software error—the failure to recognize that the probe had been programmed in both metric and English measurement units, with no means of reconciling the two—which sent the craft sailing into oblivion rather than Martian orbit. Whether the probe's loss was a true mistake or a "programmed failure," the question of whether life exists on Mars remained a closely guarded secret. *See* **Mars Observer; Mars Polar Lander**

Mars, Face on

Discovered on photographs taken by the orbiting Viking 1 spacecraft in 1976, this Martian structure, whether natural mesa or artificial construct, resembles a human face staring up into space. Wrapped in what can only be described as an ancient Egyptian-style headdress, "the Face" seems to stand guard over Mars' Cydonia plain, which appears to contain several other equally intriguing-looking features as well.

The famous Face on Mars discovered by the Viking 1 in 1976.

Widespread release of the NASA images caused a firestorm of publicity and controversy. Independent researchers became quite energetic in their pronouncements that the Face was an artificial construct, perhaps a remnant from a long-dead Martian civilization. Former science consultant Richard Hoagland believed that an entire "city," including pyramids and buildings, could be seen in the grainy images, and began campaigning for complete disclosure regarding this subject. Computer specialists Vincent DiPietro and Greg Molenaar, who reviewed the original Face image in 1979, backed up Hoagland's claim, stating that the images showed, in addition to the 1.5-mile-wide "Face," a nearly two-mile-long pyramid, which became known as the "D and M Pyramid," along with other features nicknamed "the City" and "the Fort."

NASA appeared to suffer a near-meltdown at first, trying to explain away the photographs. But many scientists rushed to their defense, claiming the low-resolution Viking images to be insufficient for making broad pronouncements about the nature of the features and often citing the believers' inability to come up with any definitive proof that the images were not just contrasts between light and shadow and the result of an observer's overactive imagination.

Despite NASA's continuing resistance to becoming embroiled in the controversy again, in 2001, with its new Mars Global Surveyor spacecraft in orbit, higher resolution photographs of the region were ordered, and NASA was able to claim that Hoagland's theories had been disproved once and for all. The new images seemed to validate that the Face was, indeed, only one of a set of natural mesa-like formations. However, NASA critics, including Hoagland, accused the space agency of tampering with the photographs and continuing to stonewall any evidence supporting the theory that a Martian civilization once existed, or even exists today. As it stands, the Face on Mars remains as controversial as it was the day it was discovered. *See* **Hoagland, Richard**

Marshall, George C.

Chief of staff of the Army in World War II, secretary of state under President Truman in 1947, and the statesman who conceived of what is now known as the Marshall Plan in 1948 to stabilize post-war Europe. Gen. Marshall was quoted by Dr. Rolf Alexander as having said, with regard to the question of UFOs:

> The United States has recovered UFOs and their occupants. The UFOs were from a different planet and they were friendly. They

have been hovering over defense facilities and airports. U.S. authorities were convinced they had nothing to fear from them. The U.S. wanted people to concentrate on the real menace, communism, and not be distracted by the visitors from space. There has actually been contact with the men in the UFOs and there have been landings.

Mars Observer

NASA's expensive and high-tech Mars probe was scheduled to reach Martian orbit in 1992. Its primary mission was to photograph objects on the planet's surface as small as a coffee table, using ultra-high-resolution cameras. Its other sophisticated sensors were expected to provide scientists with their clearest look yet at the Red Planet's geology and climate. But just minutes before the probe entered Martian orbit, NASA lost contact with it, and it was never heard from again.

Seven years later, NASA lost both its Climate Orbiter and its Polar Lander probes under equally mystifying circumstances. The space agency blamed hardware and software glitches for the failures, most of them traceable to NASA's semi-successful attempts to build "faster, cheaper, better" spacecraft during its continuing budgetary battles with the White House and Congress back at home.

But there has been speculation that the lost probes are not really lost at all—that the spacecraft are actually continuing to perform their missions of searching for life on Mars, but in a completely secret manner and for a completely hidden agenda: They are acting as the "long arm" of the federal government in its continuing efforts to hide the reality of an alien existence on earth and in space. While NASA derides these rumors, the specter of failure continues to haunt the future of Mars exploration. *See* **Hoagland, Richard; Mars Climate Orbiter; NASA**

Mars Polar Lander

Two months after the loss of the Mars Climate Orbiter in 1999, NASA's Polar Lander failed to respond to earth commands after attempting a landing near the Martian South Pole. NASA scientists said a software failure had doomed the probe to a crash landing. But many UFO researchers believed, instead, that whoever was inhabiting Mars was not ready for contact. *See* **Jet Propulsion Laboratory (JPL); Mars**

Mascons

Large local concentrations of gravity that exist just below the moon's surface. First discovered by orbiting Apollo spacecraft, the mascons' presence, some UFO researchers argue, points to the existence of large, cavernous structures that could hold some form of life. *See* **Moon, Dark Side of the**

Maury Island, Washington

One of the stranger cases in ufology occurred on July 31, 1947, when UFO debris was reportedly recovered from the Maury Island area of Puget Sound near Tacoma, Washington. Witnesses in this case were said to have been visited and threatened by "Men in Black," and the debris that was to be flown back to a U.S. Air Force base for examination never reached its destination; the plane exploded, most of the crew died, and the debris was lost.

Maussan, Jaime

A prominent news broadcaster on Mexican television's version of "60 Minutes," Maussan is also one of Mexico's leading investigators of UFOs. He has told MUFON representatives in the United States, "I have personally witnessed the sky filled with craft, while the traffic on all the freeways and roads stopped as people got out of their cars and all pointed to the sky. And not one word about these events has ever appeared in an American newspaper. I know, because I always check."

Maussan has been the lead investigator in a number of startling UFO encounter cases. He has been given dramatic film footage of silver disks hovering around one of NASA's space shuttles, and Maussan has said the disks are identical to the ones he had videotaped flying over Mexico City. *See* **Mexico City; "Tercer Milenia" ("Third Millennium")**

McBoyle, John

Station manager of the ABC affiliate radio station KSWS in Roswell who said he personally visited the crash site, then called his teletype operator, Lydia Sleppy, to tell her he had something "hot" for the network. But, according to McBoyle, just as soon as she started typing the account, the station received a teletype from the FBI ordering them to "cease transmitting." *See* **Roswell, New Mexico; Sleppy, Lydia**

McCann, Walter

On April 10, 1897, McCann photographed a large cigar-shaped craft flying over Chicago. This was part of the late-nineteenth century "airship" flap in the U.S., and a full seven years before the first dirigibles began flying. *See* **Great Airship Mystery**

McDivitt, James

A Gemini astronaut who, in orbit over the Hawaiian Islands in 1965, saw a strange-looking metallic object with long arms coming from it. Astronaut Gordon Cooper subsequently confirmed that McDivitt did take photographs of the object. *See* **Cooper, Gordon; White, Ed**

McDonald, James E.

A tireless and highly respected researcher, James E. MacDonald was a senior physicist at the Institute of Atmospheric Physics, University of Arizona, who in testimony before the House of Representatives on July 29, 1968, said:

> The type of UFO reports that are most intriguing are close-range sightings of machine-like objects of unconventional nature and unconventional performance characteristics, seen at low altitudes and sometimes even on the ground. The general public is unaware of the large number of such reports that are coming from credible witnesses. When one starts searching for such cases, their numbers are quite astonishing. Also, such sightings appear to be occurring all over the globe.

McGuire, Phyllis

Daughter of Lincoln County Sheriff George Wilcox, Mrs. McGuire was present the night Mac Brazel brought the box of Roswell saucer debris to the sheriff for safekeeping. She also preserved her mother's journal, with its recorded entry for that night. *See* **Children of Roswell**

McKittrick, Beverly

Wife of American comedian and television star Jackie Gleason, who once told friends how, in 1973, Jackie was so visibly shaken after a trip to Homestead Air Force Base to view the preserved bodies of extraterrestrial aliens, that she had to comfort him through the night when he returned. *See* **Gleason, Jackie**

McMinnville Sighting

According to a National UFO Reporting Center case brief, a January, 7, 1995, sighting took place in the skies over McMinnville, Tennessee at 7:30 P.M. when local residents from Shellsford and Irving College near McMinnville in Warren County reported to the 911 switchboard that they had witnessed strange red, green, and blue lights descending slowly and almost vertically to earth. The red light visible was the largest, with at least two smaller green lights above it. Some of the witnesses described the objects as "huge."

Witnesses also reported that one or both of the two green "clusters" of lights "exploded into fragments." Then, after the colored lights descended low enough to disappear from all observers' line of sight, an intense white flash occurred which was bright enough to be blinding. At least one person stated that he had heard the explosion concomitant to the flash. Subsequent to this sighting a local pilot took off in a private plane to see if could observe any crash site from the air, but he found nothing and returned to the airport. *See* **Davenport, Peter**

McMoneagle, Joe

Recruited as remote viewer #001 of the very black and very sensitive psychic spy unit now known as *Stargate.* McMoneagle is a respected lecturer and the author of *Mind Trek, The Ultimate Time Machine, Remote Viewing Secrets,* and T*he Stargate Chronicles. See* **Remote Viewing**

Joe McMoneagle

McMullen, Clements

An Army Air Force major general and commander of the Strategic Air Command who, on Sunday, July 6, 1947, ordered 509th Bomber Squadron commander Col. William Blanchard to retrieve some of the flying debris in Sheriff Wilcox's possession and send it immediately to Col. Thomas DuBose at Fort Worth Army Air Field. The debris was then flown from Fort Worth to Washington, where it was met personally by Gen. McMullen. *See* **Roswell, New Mexico**

Meier, Eduard "Billy"

A Swiss farmer who, in the mid-1970s, claimed contact with a race of alien beings called the Pleiadians, and who took remarkable photographs and movies of their "beam" ships. The Meier story is one of the most controversial in the annals of ufology, with some advocates calling him a prophet and others claiming that he perpetrated one of the most incredible hoaxes of all time.

The credibility of Meier's story hangs on the analysis of his photographs, stunning shots of UFOs in flight, and the information allegedly given him by the Pleiadians in the form of prophecies about planet Earth and the future of humanity. Debunkers say the photographs are merely models of spacecraft supported by wires; Meier had shot a number of these photos and films facing directly into the sun to obscure the wires supporting the models. A one-time Meier supporter allegedly told the press he saw the actual models of the spaceships in a barn; others claim that Meier built the models after he saw the spaceships.

The other controversy swirls around the prophecies that have come from the Pleiadians, the extraterrestrials who chose Meier to be their messenger to the human race. Concerned with environmentally damaging chemicals that humans had been using, and with the fallout resulting from atomic testing of the late 1940s through the 1960s, the aliens instructed Meier to contact various scientists with warnings of danger to the earth's environment.

The Pleiadians told Meier that humanity, as we understand it, originated in distant star systems. This race of humans occupied other planetary systems, including earth, and interbred with native species. They also told him that in addition to Atlantis, there was a continent of Mu, with whom the Atlanteans fought. The two continents eventually destroyed each other in this war.

Perhaps the most controversial of all Meier's statements, according to researcher Michael Horn, is Meier's assertion that he is a reincarnate. Meier's spirit, he has said, "as well as that of Enoch, Elijah, Isaiah, Jeremiah, and Mohammed, are one and the same, and that he therefore has incarnated many times in the past as a prophet."

Horn, who has written extensively on the subject for *UFO Magazine*, has argued that Meier's public predictions were based upon specific facts and information that he could not possibly have known about beforehand because of his sixth-grade education and lack of scientific expertise. In fact, Horn points out, Meier's statements, such as those concerning the destruction of the earth's ozone layer, came years before science independently corroborated his predictions. *See* **Beamships**

Men in Black (MIB)

The first widely acknowledged case of Men in Black intimidating various UFO witnesses occurred in 1953 when Albert K. Bender, editor of the magazine *Space Review* and founder of an organization called the International Flying Saucer Bureau, discovered vital information about a U.S. government cover-up of flying saucers. After being paid a visit by the MIB and told to cease his research or else, Bender was so badly scared that he officially retired from UFO investigations.

Since no agency within the United States government officially acknowledges having anything to do with such bizarre characters, UFO researchers believe they're either field investigators or operatives of the National Security Agency, or even aliens posing as government agents. In 1967, the assistant vice chief of staff of the U.S. Air Force, Lt. Gen. Hewitt T. Wheless, memoed various agencies within the Department of Defense to warn them about MIB posing as agents of the DoD. He wrote:

> Information, not verifiable, has reached Hq. USAF that persons claiming to represent the Air Force or other Defense establishments have contacted citizens who have sighted unidentified flying objects. In one reported case, an individual in civilian clothes, who represented himself as a member of NORAD, demanded and received photos belonging to a private citizen. In another, a person in an Air Force uniform approached local police and other citizens who had sighted a UFO, assembled them in a school room and told them that they did not see what they thought they saw and that they should not talk to anyone about the sighting. All military and civilian personnel, and particularly information officers and UFO investigating officers who hear of such reports, should immediately notify their local OSI offices.

See **Bender, Albert K.; Wheless, Hewitt T.**

Menzel, Donald

Reportedly a member of Majestic, astronomer Dr. Donald Menzel was known as one of the most ardent public skeptics of the existence of UFOs, challenging the reputations and credentials of anyone who publicly stated a belief in them. He was also one of the most severe critics of Immanuel Velikovsky, whose book *Worlds in Collision* challenged the tenets of established planetary science. It has been suggested by Stanton Friedman that Menzel's almost malicious debunking of UFOs was, in fact, a perfect cover for his real role as a researcher who not only knew of the existence of UFOs, but who worked for the government to cover up the truth of their existence. *See* **Majestic; Velikovsky, Immanuel**

Mexico City, Mexico

A center of UFO activity in North America. Newscaster Jaime Maussan has videotaped footage of huge flying saucers hovering over buildings and floating above the city while onlookers gasp in amazement. *See* **Maussan, Jaime**

MIB

See **Men in Black**

MILABS

See **Military Abductions**

Miley, Michael

Michael Miley

A freelance writer and researcher of transpersonal psychology and philosophy, Miley studied existentialism and radical political philosophy and became an activist in the Central American solidarity movement. A spiritual crisis in 1990 revived buried memories of childhood out-of-body, entity, and satori experiences. His focus on Western philosophy includes research into the paranormal as well as Eastern mystical studies, integrating spiritual evolutionary philosophy with the implications of the UFO phenomenon.

A contributing editor to *UFO Magazine,* Miley's work also includes comparative research in out-of-body experience, near-death experience, mystical states, reincarnation, channeling, trance and hypnosis, psychedelics, remote viewing, as well as evolutionary theory, with a view to what they reveal about human, spiritual, and extraterrestrial intelligence.

Military Abductions (MILABS)

Abductions conducted by members of the U.S. military for the purposes of either learning what happened to real alien abductees or as a part of a military or CIA "mind control" program. Cases such as these suggest that the military may be tracking alien abductees to gather intelligence on them, or that the military is cooperating with aliens in monitoring human abductees, or even that alien abductions are a cover for ongoing military/U.S. intelligence mind control/behavior control experiments to test techniques for control of key segments of a population. *See* **Leslie, Melinda; Mind Control (MK-ULTRA)**

Miller, Norman C.

Washington Bureau Chief of the *Wall Street Journal* to whom Governor Ronald Reagan of California told the story of his 1972 encounter with a UFO while flying across California. *See* **Reagan, Ronald**

Mind Control (MK-ULTRA)

A program launched by the CIA, with military participation, according to Lt. Gen. Arthur Trudeau of U.S. Army Research and Development, to influence the thought patterns and belief systems of target individuals and even entire populations.

By the mid-1950s and after the terrible discovery that during the Korean War, North Korea had brainwashed captured American troops, United States military and intelligence officials had come to realize that the Soviets and Chinese had begun massive research programs into harnessing psychic phenomena and mind-control techniques with an eye toward their possible use as an espionage, military, and political tool. American officials became determined to catch up.

From records of that period released to date, we now know that American scientists from many disciplines were asked by the CIA to identify the keys to controlling behavior and penetrating human consciousness. The researchers, as well as the Agency, were all in agreement that such tools, if harnessed and deployed with a reasonable degree of reliability, would amount to the ultimate weapon in the espionage game.

One invention that particularly interested the CIA was United States Patent Number 3,951,134, from April 20, 1976. It was an apparatus that used electromagnetic signals transmitted at various frequencies for remotely monitoring and altering brain waves to effect a desired change of behavior.

The CIA mind control programs like MK-ULTRA, its successor MK-SEARCH, and projects like Bluebird and Artichoke have all used, to one degree or another, advanced technologies such as microwave transmitters to alter perceptions and interfere with logical thinking processes, as well as more classical conditioning techniques such as hypno-programming, drugs, sensory deprivation, or electroshock therapy to create distortions of reality on test subjects—all with the aim of controlling large or small segments of a target population. While Congressional hearings during the late 1970s revealed many of these programs, to the embarrassment of the CIA and the White House, many believe the programs went "deeper black," and are still being used today for whatever aims the American government designates. *See* **Military Abductions**

Part of the protest material in response to reports of mind control testing on civilian population groups.

Ministry of Security, Russian Federation

Released a statement attesting to the fact that both the Ministry of Security and the U.S. Central Intelligence Agency and "other covert services" have enough evidence to "conclude that there is a detachment of observers from worlds traveling in near-earth orbit." *See* **Central Intelligence Agency, History of UFOs in; Russian Roswell**

Mir space station

The Russian space station, no longer in orbit, that has been the subject of speculations regarding its possible encounters with flying saucers. Russian scientists and space officials have said that for years Mir had been observed by UFOs from a near distance.

Missing Time

A term coined by researcher Budd Hopkins to describe the period of time during which abductees are taken from their homes, cars, or even public places, by aliens or military personnel posing as aliens, and subjected to either an interrogation or a battery of medical procedures. The abductors usually erase the abductees' memories or implant false memories to cover up what has taken place. Abductees usually find themselves unable to account for this period of time during which they were under the control of their abductors. Regression therapy, as practiced by Hopkins and others, has enabled some abductees to recall the blocked or erased memories. *See* **Hopkins, Budd; Mack, Jonathan**

Mitchell, Edgar

Although Apollo 14 astronaut Edgar Mitchell has said that he's had no personal experience with extraterrestrials or UFOs, he's been very outspoken in his belief of their existence. On NBC's "Dateline" in 1996, for example, Dr. Mitchell stated that he had spoken with people who have first-hand experience with UFOs: "Their evidence is very strong, and large portions of it are classified by governments." In 1991 Mitchell said, "I do believe there's a lot more known about extraterrestrials than is available to the public right now, and that's been true for a long time. It's a long story, which goes back to World War II when all that happened and is highly classified stuff."

MK-ULTRA

See Mind Control

Mogul

See Flight 9; Projects, Government UFO

Molecular Computing (DNA Computer)

A computer that operates on the basis of replicating DNA and enzymes, which stimulates the replication and fuels the operation as well. The enzyme FokI breaks bonds in the DNA double helix, causing the release of enough energy for the system to be self-sufficient so that the computing process powers its own operation. Less than ten years ago, scientists in California used test-tube-stored DNA to solve simple mathematical problems, but the process had to be sustained by an outside power source. Recent breakthroughs allow the DNA molecules themselves to power the process; thus the entire computer carries its own energy source.

Some UFO researchers believe that DNA computers are a part of the genetic makeup of extraterrestrials and control the adaptation of cloned EBEs to native environments. This allows the onboard spacecraft's computer to manufacture or modify clone scout-grays specific to a host-planet's environment so the scout can perform experiments, survey the environment, and interact with the host-planet's life forms. This is how the known galaxy might have been explored and seeded billions of years ago by an advanced species. *See* **Crick, Francis**

Monroe, Marilyn

The movie star whose reported love affairs with President John F. Kennedy and his brother Robert fueled speculation regarding her knowledge of highly classified information about UFOs. At a Reno casino, for example, an intoxicated Monroe announced that she was tired of being passed around by the Kennedy brothers and was going to get back at them by telling their secrets. She mentioned a secret U.S. air base that had "things from outer space" and "little men" recovered at a crash in the desert. Although Monroe's death was officially classified as a suicide, there were many doubts about what really happened. A number of her biographers claimed that her death was more likely a homicide. *See* **Kennedy, Robert F.**

Montague, Robert M.

Head of the Armed Forces Special Weapons Center, member of MJ-12. *See* **Majestic**

Moon, Dark Side of the

Although "the dark side of the moon" refers to the lunar hemisphere that is always turned away from the earth, it has also come to be a metaphor for what some people believe are lunar mysteries that still persist even after the manned Apollo lunar program.

The story of the exploration of the moon is one of fits and starts, of billions of dollars spent and grand explorations undertaken and then literally abandoned. Thousands of photographs later, after scores of unmanned probes and six manned voyages to its surface, we still don't have a handle on what our closest celestial neighbor is or what it contains. Were there aliens already on the moon to greet the first Apollo astronauts? Are there artifacts there from another species? Did an alien spaceship fire upon the astronauts, which is what one

An area on lunar surface where astronomers have observed anomalous structures.

radio transmission seemed to indicate? And did NASA's study of five hundred years of lunar anomalies—Report R 277—point to the existence of a lunar colony populated by extraterrestrials?

On December 29, 1986, Neil Armstrong, the first man to step onto the moon, told author Timothy Good, "There were no objects reported, found, or seen on Apollo 11 or any other Apollo flight other than of natural origin. All observations on all Apollo flights were fully reported to the public."

However, rumors have persisted for years that the astronauts, most of whom are military officers susceptible to being silenced under orders, may have observed more than just rocks and dust. According to transcripts of the technical debriefing following the Apollo 11 mission, astronauts Armstrong, Buzz Aldrin, and Michael Collins told of an encounter with a large cylindrical UFO even before reaching the moon. Aldrin said, "The first unusual thing that we saw I guess was pretty close to the moon. It had a sizable dimension to it." Aldrin said the Apollo crew at first thought the object was part of their booster rocket, but added, "We called the ground and were told the S-IVB was 6,000 miles away." Aldrin described the UFO as a cylinder, while Armstrong said it was "really two rings. Two connected rings." Collins also said it appeared to be a hollow cylinder that was tumbling. He added, "But then you could change the focus on the sextant and it would be replaced by this open-book shape. It was really weird."

One account of how the astronauts were muzzled came from an ex-NASA employee named Otto Binder, who accused his former employer of censoring transmissions from Apollo 11 and other missions by switching to radio channels unknown to the public. He claimed that unnamed ham radio enthusiasts monitored Apollo 11 transmissions after the astronauts landed in the Sea of Tranquillity and overheard one of the astronauts exclaim, "These babies are huge, sir ... enormous ... Oh, God, you wouldn't believe it! I'm telling you there are other spacecraft out there ... lined up on the far side of the crater edge ... they're on the moon watching us."

A variation on this story was repeated by Maurice Chatelain in his book *Our Ancestors Came from Outer Space*. Chatelain wrote, "When Apollo 11 made the first landing on the Sea of Tranquillity, only moments before Armstrong stepped down the ladder to set foot on the surface, two UFOs hovered overhead." He added, "The astronauts saw things during their missions that could not be discussed with anybody outside NASA. It is very difficult to obtain any specific information

Neil Armstrong's one small step for man onto the lunar surface on July 20, 1969.

from NASA, which exercises strict control over any disclosure."

Chatelain also acknowledged that NASA had the capability to hide secret Apollo transmissions among a wide variety of radio channels. "For instance, there were seven channels to feed medical information about the physical condition of the astronauts, and nine to retransmit the stored telemetry data from the passage behind the moon that could not be beamed directly," he stated. All such communications could easily be hidden from the public.

In 1995, *UFO Magazine* Research Director Don Ecker described not only mysterious lights on the moon, but actual structures that seemed to be artificial in nature. There was "the Bridge," a structure that looked as if it were straddling a crater. British astronomer Dr. H. P. Wilkens said of it: "It looks artificial. It's almost incredible that such a thing could have formed in the first instance, or if it was formed, could have lasted during the ages in which the moon was in existence."

Other structures include "the Shard," a spindle-shaped object that stands a mile and a half high and is composed of regular geometric patterns of the type that do not occur as the result of natural formation. Then there's "the Tower," a five-mile-high structure whose top has a "very ordered cubic geometry," according to geologist Dr. Bruce Cornett. Like the Shard, the Tower looks as though it has been constructed by an intelligent species, possibly an alien race. *See* **Apollo Program; Wolfe, Karl**

NASA's Lunar Lander.

Moonbase, Project Horizon

Buried away among classified documents was a set of plans once so top secret that not even astronaut and former Senator John Glenn of Ohio knew of its existence. This U.S. Army program was known as Project Horizon, the plan for a fortified moon base, fully endorsed by the Director of Army Research and Development, Lt. Gen. Arthur Trudeau. It called for a single Army command to build and occupy the base so as to forestall the moon's occupation by Soviet or other hostile forces.

The cover of the declassified government project to build a military base on the moon.

Project Horizon was never implemented. In fact, one of the goals of the Apollo lunar missions was to establish a similar base, which was never completed either. Some UFO researchers have speculated that the Apollo missions drew enough attention—even hostile intentions—from aliens that NASA was literally scared off the moon.

Moore, Patrick

A member of the British Astronomical Association who, in confirming the existence of "the bridge" on the moon's surface, said that the bridge had appeared almost overnight. *See* **Moon, Dark Side of the; O'Neill, John J; Wilkins, H. P**

Moreno, Sanchez

Argentinean naval captain Moreno, the Naval Air Station Comandante Espora, said in a 1979 speech in New York:

> The unidentified flying objects do exist. Their presence and intelligent displacement in the Argentine airspace has been proven. Their nature and origin is unknown and no judgement is made about them.
>
> Between 1950 and 1965, personnel of Argentina's Navy alone made twenty-two sightings of unidentified flying objects that were not airplanes, satellites, weather balloons, or any type of known vehicles. These twenty-two cases served as precedents for intensifying that investigation of the subject by the Navy. In the past 2

years, nine incidents have been recorded that are being studied by Captain Pagani and a team of military and civilian scientists and collaborators.

Morningsky, Robert

A speaker on UFO-related topics who revealed that in 1947 his grandfather and others found a downed flying saucer in New Mexico with a live alien on board.

Mothman

See **Keel, John**

Mount Palomar Observatory

Site of an astronomical facility in California. Also the location where 1950s UFO contactee George Adamski reported that he saw flying saucers on over two hundred occasions.

Mount Popocatepetl

An active volcano in Mexico where many flying saucers have been observed. Mount Popocatepetl is reported to have the strongest electromagnetic fields of any location on earth.

Musgrave, Story

A space shuttle astronaut and spacewalker, Musgrave once told the *Houston Post* that he tries to communicate with extraterrestrials, "when I'm circling around out there." He tries to telepathically convince them to "come down here and get me."

Mutual UFO Network (MUFON)

The Mutual UFO Network was founded in 1969 by Walt Andrus for the purpose of conducting independent, standardized, and objective investigations into the sightings of UFOs. Andrus retired in 2001. The current president is John Schuessler. *See* **Andrus, Walt; Schuessler, John**

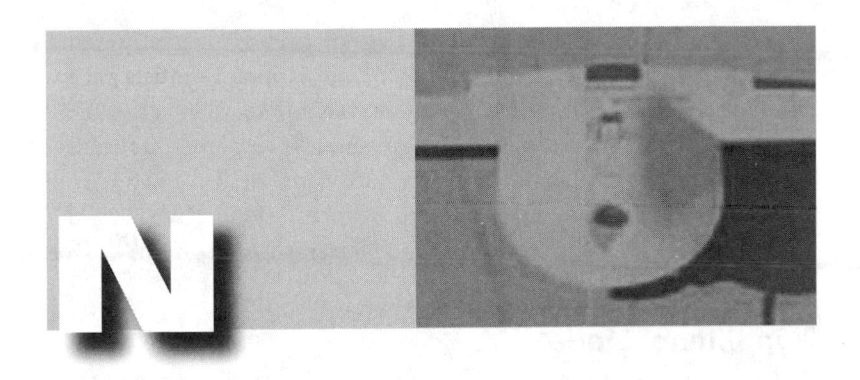

Naiman, Robert

The psychiatrist whose help was enlisted by abductee researcher Budd Hopkins to regress a group of people who claimed to have had strange experiences during a hike in the woods, where they said they were kidnapped by strange-looking creatures and subjected to a variety of experiments. *See* **Hopkins, Budd**

Nanotechnology and UFOs

Nanotechnology is the extreme miniaturization of electronic devices to the cellular or atomic level by treating atoms discretely, in the same way that computers treat bits of information. Within the past decade scientists have developed tools capable of manipulating individual atoms.

Nanotechnology is molecular manufacturing, building things one atom or molecule at a time with programmed nanoscopic robot arms. Thus, nanotechnicians are able to manipulate atoms individually and place them exactly where needed to produce the desired structure.

The goal of early nanotechnology is to produce the first nano-sized robot arm capable of manipulating atoms and molecules into a useful product. The arm could make an almost unlimited number of copies of itself so as to array trillions of assemblers controlled by nano-supercomputers working in parallel, thus assembling objects quickly. The hoped-for result would be microscopic computers working inside the human organism as artificial immune systems. These machines could manipulate DNA itself to repair cells and stop or reverse aging.

This process, UFO researchers believe, is part of an ongoing and self-sustaining project of various alien races to explore the galaxy, populate various planets the collective nanobots themselves may select, and assemble biologically engineered "rovers" to conduct experiments on those planets and their native species. This, U.S. Army pathologists at Walter Reed Hospital believed, might have been the type of species that was recovered from the crash at Roswell. *See* **Roswell, New Mexico**

Napolitano, Linda

The real name of the abductee whom Budd Hopkins referred to as *Linda Cortile* in his story of the "Brooklyn Bridge Incident," which took place on November 30, 1989 in lower Manhattan. Napolitano/Cortile was floated out of her apartment and into a UFO mothership, which then dove into the East River. *See* **Brooklyn Bridge Abduction Case; Cortile, Linda**

NASA

See **National Aeronautics and Space Administration**

National Aeronautics and Space Administration (NASA)

The federal agency in charge of America's civilian space program. NASA was created during the Eisenhower administration as an open agency with a civilian agenda, but it has increasingly conducted military projects and has reportedly gone "black" since its formation.

One of NASA's greatest challenges, according to many UFO researchers, has been to keep from the American people the truth about encounters with alien spacecraft orbiting the earth and alien bases on

The Tier-3 Unmanned Aerial Vehicle, 1997. UAVs played a vital role in Afghanistan and Iraq in 2002 and 2003.

the moon. To this end, NASA has developed a high-tech sanitizing protocol to bounce transmissions from the astronauts off a number of satellites so as to prevent private parties on earth from receiving uncensored messages.

NASA has also instructed its astronauts not to transmit messages about UFO encounters or any other type of alien manifestation they may observe or come in contact with while in space. However, over the years, a number of astronauts have eventually gone public about their encounters.

A number of transmissions between astronauts and NASA controllers regarding the presence of alien spaceships have been picked up by private parties and released to the public. *See* **Hoagland, Richard**

National Aeronautics and Space Administration Technical Report # R-277

The space agency's study of five hundred years of lunar anomalies, released before the Apollo astronauts reached the moon in 1969. It is known as the "Chronological Catalog of Reported Lunar Events." Some of the observations listed in the report are quite striking, including those referring to clouds of gas moving across the lunar surface and formations of lights hovering above the moon. The report is widely available at a number of UFO sites on the Internet.

National Aerospace Council, Indonesia

In a letter published in *UFO News,* the Secretary of the National Aerospace Council, Air Commander J. Salutun, said that he was convinced that nations needed to study the "UFO problem seriously for reasons of sociology, technology, and security."

National Centre for Space Studies, France

Reports coming into this organization were the subject of a radio interview by the French minister of defense in which he said, "If listeners could see for themselves the mass of reports coming in from the airborne *gendarmerie,* from the mobile *gendarmerie,* and from the *gendarmerie* charged with the job of conducting investigations . . . they would see that it is all pretty disturbing."

National Institute for Discovery Science (NIDS)

A privately funded, private-sector research organization focusing on empirical and hypothesis-based scientific exploration of aerial phenomena to expand conventional knowledge. NIDS is interested in phenomena that appear to be under intelligent guidance and in the possible habitation of space by intelligent life.

National Investigations Committee on Aerial Phenomena (NICAP)

National Investigations Committee on Aeriel Phenomena, headed by Donald Keyhoe in the 1950s. This organization was tasked to study reports of aerial anomalies that were being increasingly turned in to authorities since the highly reported crash of an object in Roswell.

NICAP former investigators. Rear left to right: Jan Aldrich, Richard Hall, Bernard Haugen, Don Berliner, Nick Summers. Front left to right: E.R. Sabo, Melvin Podell, Larry Bryant.

Founded by Tommy Townsend Brown and later run by Maj. Keyhoe, NICAP investigated UFO sighting reports from 1956 to 1973, often coming up with information that was of great interest to the CIA and other intelligence services. Roscoe Hillenkoetter, one of the original directors of the CIA, was on NICAP's board. Karl Pflock, former CIA officer and now an official debunker of the existence of UFOs was also an officer of NICAP. NICAP's files were given to J. Allen Hynek after the body was dissolved and they later became part of the information database of the Center for UFO Studies (CUFOS).

The CIA's interest in NICAP stemmed from the organization's efforts in reporting the true facts behind the UFO phenomenon at a time when the U.S. Air Force was running Project Blue Book, set up to debunk the UFO reality. *See* **Center for UFO Studies (CUFOS); Hynek, J. Allen; Keyhoe, Donald**

National Security Agency (NSA)

A covert intelligence-gathering agency involved in the worldwide monitoring of communications, and harvesting those communications for their intelligence value relative to the national security of the United States. The NSA operates land, sea, air, and space-based sensors to monitor civilian as well as military and diplomatic communications. NSA agents have sometimes been identified as "Men in Black" investigating reports of UFO encounters.

The NSA, an operational arm of the National Security Council, tasked the CIA with gathering as much information on UFOs as possible and as recently as 1975 memoed the CIA that it was imperative to study UFOs so that they don't "confuse" U.S. early-warning defense systems in the event of an attack by the Soviet Union. *See* **National Security Council (NSC); North American Air Defense Command (NORAD)**

National Security Council (NSC)

The NSC is, effectively, the fourth branch of the United States government with powers that, in some instances of national security, supersede the powers of the president of the United States, who also sits on the Council. NSC directives and regulations, for example, might have precluded President Jimmy Carter from releasing any information on UFOs to the American public even after he promised to do so. *See* **Central Intelligence Agency, History of UFOs in; Men in Black (MIB); National Security Agency (NSA)**

National UFO Reporting Center (NUFORC)

Located in Seattle, Washington, and founded in 1974 by UFO investigator Robert J. Gribble, the Center is now run by Peter Davenport. It records and corroborates reports from individuals who have witnessed unusual, possibly UFO-related events. Calls regarding recent sightings are welcomed by the hotline at (206) 722-3000. The hotline is staffed 24 hours a day. *See* **Davenport, Peter**

Native Americans

Long before the U.S. military and the CIA began their experiments in remote viewing, Native American shamans, or "spirit-walkers," undertook vision quests. These psychic expeditions were out-of-body experiences and the precursor to the more modern remote-viewing experience. On these out-of-body excursions, dream-walkers or spirit-walkers encountered avatars of others like them in the extra-dimensional realm. It has been said that Native American spirit-walkers also encountered alien species in this realm as well, and perhaps were the first explorers to make extraterrestrial contact. *See* **Remote Viewing**

NATO

See **North Atlantic Treaty Organization**

Nazis and UFOs

One of the ongoing debates among UFO researchers has been the extent to which, if at all, the Nazis were responsible for developing advanced propulsion technologies during World War II that are, in fact, the basis for the modern UFO phenomenon. A subcategory to this raises the possibility that the Nazis had recovered downed extraterrestrial flying saucers and tried to reverse-engineer that technology into weapons. There are those within the UFO community who stoutly defend and deride these theories.

One such theory suggests that the Germans acquired UFO technology as far back as the 1930s but pressure from the German high command to pursue the war put research and development on hold, and the weapons that could have been reverse-engineered from UFO technology were never completed.

Those projects, whatever they were, eventually found their way into American and Russian laboratories after the war. George Earley's recent *UFO Magazine* review of Henry Stevens' *Hitler's Flying Saucers* is instructive in that it poses the question: If America actually captured and put to use German technology after the war, why hasn't that technology been used in Korea, Vietnam, or even more recently, the Persian Gulf?

The truth is, Earley writes, that Hitler probably didn't have extraterrestrial saucers—just a few experimental home-grown jet-powered ones—and that whatever is flying around in our skies today, it is probably not the result of German engineering.

Near-Death Experience (NDE)

In a near-death experience, there is a cascade of chemicals released in the brain that trigger a complete holographic memory of one's life, which is then discharged into the person's consciousness. Long buried or hidden memories come to the fore, and the person relives them in an instant as the body begins to shut down and the oxygen-deprived brain begins to access core thoughts and emotions without any logical censoring.

Those who study NDEs, particularly in children, talk about a commonality of shared visions as the person approaches death, such as a tunnel of light, the presence of close relatives who have died, and a sense of acceptance of the inevitable. Those who have returned from NDEs have said that they became remarkably clairvoyant during this state and, though apparently comatose, reportedly were able to describe the events taking place around them with stunning detail.

Some UFO researchers have said that extraterrestrials have been able to create the symptoms of near-death in some of their abducted subjects to study how human consciousness operates and, more importantly, to see if humans actually ascend to another plane of existence.

Newark Airport

The site of a remarkable encounter between commercial jets and a UFO on November 17, 1997, all of which was picked up by a ham radio operator named John Gonzales, (call sign NZIXW), sometime before midnight. It is believed by many UFO researchers that this sort of encounter is quite typical, and although these sightings by pilots are frequent, they are almost never reported. The crew members of two other aircraft in the area witnessed the sighting and joined in the radio conversation. See **Gonzales, John**

Newton, Irving

A young meteorology officer at the U.S. Army's Fort Worth base, ordered by Gen. Roger Ramey to confirm the identity of wreckage spread out before assembled reporters as that of a weather balloon. Ramey supposedly made the switch from the Roswell debris to the weather balloon after Maj. Jesse Marcel and Col. Thomas DuBose had left for Washington. The photo of Irving Newton and the weather balloon became one of the famous cover-up pictures. *See* **DuBose, Thomas J.; Ramey, Roger; Roswell, New Mexico**

Newton, Silas

In March of 1950, Silas Newton was reported to have given a lecture at the University of Denver concerning government retrievals of crashed UFOs. Among the list of crashes from which the government supposedly retrieved debris of some significance were those in the American Southwest in 1947 and in Mexico in 1948.

NICAP

See **National Investigation Committee on Aerial Phenomena**

Nicks, Ronald

Nicks was the geologist who examined photos taken during the first seven minutes of the Mars Pathfinder's transmission in 1997 and said that the unaltered high-resolution photos, which NASA no longer makes available to the public, reveal the Martian rocks in the photos are not rocks at all but actually artifacts resembling tubes, wheels, and gears.

NIDS

See **National Institute for Discovery Science**

Nixon, Richard

Thirty-seventh president of the United States, who reportedly escorted comedian Jackie Gleason to Homestead Air Force Base in Florida, where the comedian was allowed to gaze at mangled alien bodies stored in freezer cases, along with remnants of a saucer. This so changed Gleason's life, his widow once said, that he was unable to eat or get a solid night's sleep for weeks afterward. *See* **Gleason, Jackie**

NOAA Satellites

Weather satellites that have reportedly taken photographs and recorded infrared signatures of flying saucers in earth's atmosphere. *See* **Fastwalkers**

Noory, George

George Noory

George Noory, host of Premiere Radio's "Coast to Coast AM," is the top-rated nighttime radio talk show host in the U.S. with a nightly audience of over ten million listeners, the largest late-night audience in the country. At age 28 Noory was the youngest major news market news director in the country when he was at KMSP-TV in Minneapolis. But it was on his late-night radio program on KTRS in St. Louis, where he was known as *The Nighthawk,* that George's penchant for covering paranormal topics with enthusiasm and skill was honed. Noory took over the show when its then-current host Art Bell retired. *See* **Bell, Art; Lonehood, Lex**

NORAD

See North American Air Defense Command

North American Air Defense Command (NORAD)

Located in Colorado, this is the continent's primary early warning and command-and-control facility. NORAD manages a sophisticated array of satellites and radar stations to track air and space traffic and to warn of an enemy attack.

NORAD's sensors have also reportedly picked up large numbers of otherwise unidentified aircraft over the years, all of which display unusual flight characteristics, such as hyper-speed and maneuverability. These objects have become so common that NORAD does not allow them to trigger defense systems anymore; command officers merely note them in their reports. *See* **Fastwalkers; Katchen, Lee**

Nordics

Referred to by abduction researchers as the *Nordic Ones,* these beings look identical to humans and are here in great numbers, walking, living, and working among us without being recognized as extraterrestrials. They are over six feet tall and both males and females of the species have long blond hair.

Nordics are intrigued by humanity's diversity of spirit, and they are also amazed at humanity's fear of its own diversity. Rather than valuing it, the Nordics say, humans condemn it and teach their young to react violently towards people of other races. It is this violent part of human nature that has alarmed beings from other worlds, the Nordics have told their contactees, and it is they who fear us and our capability to inflict violence on other planets in the galaxy. Nordics have said that human beings are being kept within a tightly circumscribed perimeter around this planet in order to contain that violence, with the full knowledge of their governments, who are aware of the role extraterrestrials are playing on the Earth.

North Atlantic Treaty Organization (NATO)

A military pact formed by the Allied powers after World War II as a bulwark against the expansionist nature of the Communist Warsaw Pact. NATO airbases, such as RAF Bentwaters in the U.K., were the sites of bizarre UFO incidents and encounters. In Belgium, another

NATO member, there were many sightings of flying triangles, and Belgian aircraft scrambling to intercept them were unable to match their speed, maneuverability, or climb rate. NATO reports have been a prime source of information regarding the interactions between Allied air forces and UFOs. *See* **Bentwaters Case**

North Pole

The North Pole is the assumed "attack route" to be used by U.S. and Soviet/Russian missiles and bombers in case of war, since it represents the shortest distance between the two continents. In the autumn of 1960 all the nuclear bombers at Travis Air Force Base were put on red alert for an attack against the Soviet Union after radar detected targets flying towards North America over the pole. The targets suddenly disappeared from radar, and were later determined to be radar reflections from the moon.

The North Pole is also the location for the reported "Hole in the Pole," a portal to "inner earth" discovered during an American polar expedition in 1912 led by Admiral Richard Byrd, where large subterranean caverns were said to house fleets of alien flying saucers and a race of non-human beings. Though a product of misinterpreted photographs and embellishments by fiction writers, the North Pole has, in fact, been the site of numerous UFO sightings over the years.

Northrop, Jack

Founder of the Northrop Corporation, Jack Northrop reportedly had direct access to the data acquired from the Roswell spacecraft. Northrop spent the 1940s working on his design for a flying wing and by 1948 had developed the YB-49 bomber—which supposedly bore a remarkable resemblance to the delta-shaped craft that was reported to have crashed at Roswell in 1947. This has led some researchers to believe there is a direct link between the flying-wing designs of the Roswell craft, the flying triangles that were seen over Belgium in 1989–90 and in New York's Hudson Valley in the mid-1980s, and today's B-2 Stealth bomber. *See* **Roswell Spacecraft**

NSA

See **National Security Agency**

NSC

See **National Security Council**

NUFORC

See National UFO Reporting Center

Nurjadin, Rosemin

An air marshal and commander in chief of the Indonesian Air Force who said in a 1967 letter:

> UFOs sighted over Indonesia are identical with those sighted in other countries. Sometimes they pose a problem for our air defence and once we were obliged to open fire on them.

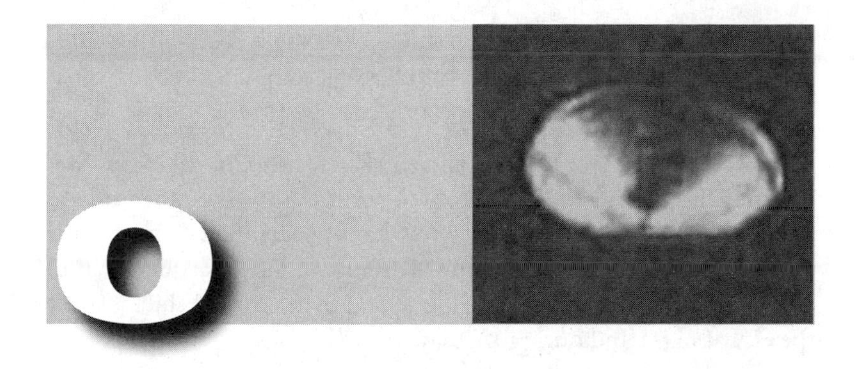

Obelisks, Lunar

Large, sharply pointed objects, some taller than the Washington Monument, photographed and observed on the moon's surface by astronomers and astronauts. Some UFO researchers have said that the "obelisks" cannot possibly be a naturally occurring formation and thus point to the possibility that alien cultures have constructed them for their own purposes. *See* **Moon, Dark Side of the**

Oberth, Hermann

German rocket pioneer who promoted space travel in the 1930s. Oberth worked with Wehrner von Braun at the V2 rocket complex near Peenemünde, and after the war joined von Braun at the U.S. Army Ballistic Missile Agency in Huntsville, Alabama. Dr. Oberth also worked as a consultant for NASA.

In an article entitled "Flying Saucers Come from a Distant World" in the October 24, 1954 issue of *American Weekly*, Dr. Oberth wrote: "It's my thesis that flying saucers are real and that they are spaceships from another solar system. I think that they possibly are manned by intelligent observers who are members of a race that may have been investigating our earth for centuries. I think that they possibly have been sent out to conduct systematic, long-range investigations, first of men, animals, vegetation, and more recently of atomic centers, armaments, and centers of armament production."

Oberth was also quoted in the same article as having said, "We cannot take the credit for our record advancement in certain scientific fields alone. We have been helped [by] the people of other worlds." *See* ***Fate*** **Magazine; Zero-Point Propulsion**

Obsequens, Julius

A writer and chronicler of ancient Rome who, in 393 CE during the reign of Theodosius, wrote about the huge luminous sphere accompanied by many smaller ones that appeared over grain fields outside of the city, panicking the citizenry. Julius Obsequens wrote that people had seen "burning globes" and flying "round shields" over the city at night and during the day.

Oeschler, Robert

A NASA mission specialist from 1974-1977, Oeschler has said he was "convinced" that the U.S. government "does indeed have craft that meet the description of many UFO reports," based upon conversations he has had with members of the U.S. intelligence community. *See* **National Aeronautics and Space Administration (NASA)**

O'Leary, Brian

A former astronaut who told the International Forum on New Science at Fort Collins, Colorado, in September of 1994, that "We have contact with alien cultures." *See* **Apollo Program; National Aeronautics and Space Administration (NASA)**

Oliphant, Ted

UFO investigator and former police officer who asked debunker Phil Klass what he really thought about the existence of UFOs. Klass allegedly replied that he supposed another civilization sends a probe to Earth every hundred years. *See* **Debunkers; Klass, Philip**

Oliveira, Joao Adil

Chief of the Brazilian Air Force General Staff Information Service, who, in a briefing to the Army War College in 1954, said:

> The problem of flying discs has polarized the attention of the whole world, but it's [a] serious [subject] and deserves to be treated seriously. Almost all the governments of the great powers are interested in it, dealing with it in a serious and confidential manner, due to its military interest.
>
> And in an interview in *O Globo,* the brigadier general said, "It is impossible to deny any more the existence of flying saucers at the

present time. The flying saucer is not a ghost from another dimension or a mysterious dragon. It is a fact confirmed by material evidence. There are thousands of documents, photos, and sighting reports demonstrating its existence.

O'Neill, John J.

The *New York Herald Tribune* science editor who, in 1954, saw a twelve-mile-long "bridge" on the moon. A month later, Dr. H. P. Wilkens, a British astronomer, came forward and said that he, too, had seen the bridge. The bridge was subsequently confirmed by Patrick Moore, a member of the British Astronomical Association. *See* **Moore, Patrick; Wilkins, H. P.**

160th SOAR Division

Made up of some of the most skilled aviators and support soldiers in the U.S. Army, the 160th's mission is to fly nighttime special operations assault raids deep into enemy territory—hence the unit's nickname, "Night Stalkers."

The SOAR division has also conducted war-game exercises in various U.S. cities, spurring conspiracy/New World Order theorists to fears of a military takeover. In UFO circles, however, the SOAR division is most closely related to the appearance of black helicopters in the vicinity of downed UFOs or UFO retrievals. *See* **Black Helicopters**

Open Skies Treaty

As the United States and Soviet Union faced off against each other in the Cold War, President Eisenhower, in order to stave off what could be a fatal mistake if either power believed the other was preparing a nuclear first strike, agreed to engage in a project called *Open Skies* in which each country would be allowed to surveil the other to make sure neither side was preparing for war. Some UFO researchers have complained that as a result of open skies, even hostile governments know more about the goings on at Area 51 than the American people do, and the American people are the ones who are paying for it. *See* **Area 51**

Operation Bluefly

An otherwise secret government project tasked with cataloging all objects that fall to earth from space. Although its "all objects" mandate includes meteorites that make it through the atmosphere, parts of satellites and rockets and other routine debris, it most specifically refers to UFO retrieval. Reportedly, there is a huge catalog on file of all the UFO debris and craft that have been retrieved by the United States since 1947. *See* **Stone, Clifford**

Operation Moonblink

A NASA study of anomalous lights on the moon. The space agency commissioned observatories from around the world to observe and photograph the moon, and within months there were more than twenty-eight documented lunar "events." Photographs of these events have never been released to the public. *See* **Moon, Dark Side of the**

Operation Moondust

The actual physical retrieval of debris from space, most specifically that from UFOs. Reportedly, the recovery operations are very extensive and elaborate, cloaked as they are in complete secrecy, with specially designed aircraft and trained personnel. UFO researcher Clifford Stone, has said that his military unit was part of a nuclear/biological/chemical/hazardous materials unit that closed off a crash area under the guise of toxic danger, enabling the unit to recover the remains of a UFO. *See* **Stone, Clifford**

Orchard Park POW camp

A prison camp that housed German POWs during World War II, located just outside of Roswell, New Mexico. Roswell crash witness Frankie Rowe testified that a military officer threatened her and her family after the Roswell crash, warning that they would be incarcerated at Orchard Park if she ever told anyone about the mysterious metallic fabric from the spacecraft that she had handled. *See* **Children of Roswell; Rowe, Frankie**

OVNI, The French Report

The French Institute for Higher Defense Studies report on OVNI ("flying objects not identified"). Made public in 1999, it revealed not only high-level concern for, and formerly classified information about, UFOs, but called for a methodical approach to the problems of French national security posed by these unidentified flying objects whose hostile intent could not be ruled out.

Written by an independent group of researchers and authors, called COMETA (Committee for In-Depth Studies) who worked with the Institute of Higher Studies for National Defense, the report demands a defense ministry study at the highest levels into how the country should prepare for the possibility of an invasion by extraterrestrials, the discovery of a base of extraterrestrials on earth, or the public announcement by aliens from another planet that they've arrived on earth and seek contact with humans.

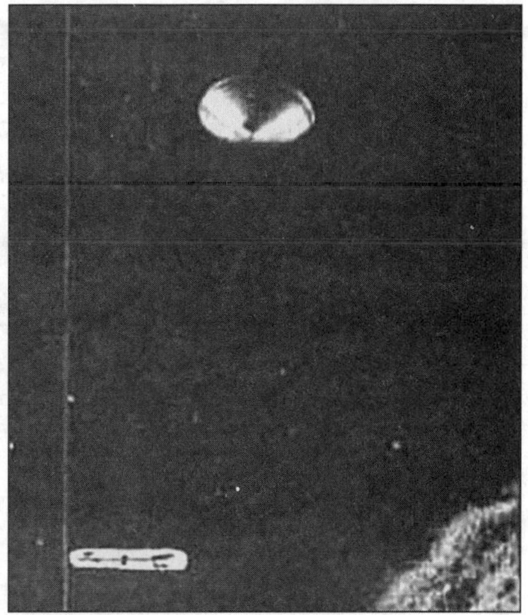

One of the few UFO photos from Project Blue Book not explained away as a natural phenomenon or a hoax.

Divided into three parts, the report begins with factual and eyewitness testimonies from witnesses to UFO encounters, assesses the current state of knowledge of UFO phenomena, and evaluates issues of national defense in terms of the potential capabilities of UFOs, especially if those UFOs are extraterrestrial in origin. In the final section of recommendations, the authors propose strategies to pursue organized studies into UFO phenomena and ways to amalgamate research into organizations that supersede traditional disciplines and national interests so as to prepare for eventual contact with intelligent beings who, with a quasi-certainty, are extraterrestrial.

Two of the primary questions the report asks are, "What situations ought we to prepare ourselves for? What strategies must we develop should extraterrestrials make contact with us?"

The report looks at the following scenarios:

1. A UFO appears, and its extraterrestrial passengers or flight crew express a desire to establish a public and peaceful contact with Earth's official representatives.
2. The accidental or deliberate discovery of an alien base either in France or somewhere else in Europe where the

aliens may or may not have established a friendly presence.

3. A general invasion by extraterrestrials or a local attack at some strategic location.
4. A manipulation or disinformation campaign aimed at destabilizing national governments.

Perhaps this latter strategy has already begun and both the secrecy and proliferation of reports are the substance of a campaign designed to weaken governments while preparing earth's population for an alien presence. Perhaps that presence has already been established and the aliens themselves or their confederates on earth are manipulating a vast public information and media campaign. It that's the case, who's left to determine what's true and what's false?

Oz Effect

The physical and neuropsychological phenomenon associated with encounters with UFOs or extraterrestrials in which the intense amount of static electricity or electromagnetic activity produced by the craft or being slows down the physical and mental processes of the human witness. The witness, unable to move as easily as he or she is accustomed to and processing sensory data more slowly, experiences what many psychologists call an "associative dysfunction" in which the person believes he or she is in a state of unreality. This feeling of unreality, of a disconnect with his or her physical surroundings, is actually a reaction to the intense electromagnetic or static field generated by the object. Because of the witness' sense of unreality, the phenomenon was dubbed *the Oz Effect* by UFO researchers. Most notably, the Oz Effect was cited as the reason for the heightened states of anxiety of United States Air Force personnel approaching the object in Rendlesham Forest during the RAF Bentwaters incident in 1980. *See* **Bentwaters Case**

Padrón, Francisco

A witness to the Canary Islands UFO on June 22, 1976. Dr. Padrón filed a formal deposition as to what he and other residents of the village of La Rosas had seen, in which he swore:

> It was made of a totally transparent and crystalline-like material, since it was possible to see through it the stars in the sky; it had an electric blue color but tenuous, without dazzling. It had a radius of about 30 meters, and in the lower third of the sphere you could see a platform of aluminum-like color as if made of metal, and three large consoles. At each side of the center, there were two huge figures of 2.5 to 3 meters tall [10 feet], dressed entirely in red and facing each other in such a way that I always saw their profile.
>
> Then I observed that some kind of bluish smoke was coming out from a semi-transparent central tube in the sphere, covering the periphery of the sphere's interior without leaking outside at any moment. Then the sphere began to grow and grow until it became huge, like a 20-story house, but the platform and the crew remained the same size. It rose slowly and majestically, moving slowly toward Tenerife. Suddenly, it reached enormous speed like none I ever saw in an airplane. Then, it disappeared in the direction of Tenerife.

See **Canary Islands UFO Case**

Pathfinder Rover

The Mars Pathfinder Rover allegedly lost power and stopped transmitting from Mars in March, 1998. At least that's what NASA said. But the little wobbly remote controlled robot vehicle that was nicknamed "Yogi" by its controllers at NASA was the subject of controversy well before it stopped transmitting. See Mars; **National Aeronautics and Space Administration (NASA)**

Parker, Calvin

UFO abductee, along with Charles Hickson, whose experiences made national news and became known as the "Pascagoula Affair." Parker claimed that he and Hickson, both workers at the local shipyard, were fishing on a dock when a UFO suddenly appeared behind them. Three robot-like beings emerged from the craft, "floated" through the air, and took both Hickson and Parker back inside the UFO. An examination was performed on both men, who were later returned to the dock.

The national media, especially the tabloids, enjoyed playing up the abduction story of two local "hicks" from rural Mississippi, but the elements of this particular story have survived media derision because they comport with the basis elements of the abduction experience, including the sighting of a strange object, a dreamlike experience involving strange-looking humanoid creatures, dreamlike memories of being inside an unfamiliar setting, missing time, and a return to reality in a state of extreme emotional discomfort or panic. *See* **Pascagoula Affair**

Pascagoula Affair

Site of the October 1973 abduction of Charles Hickson and Calvin Parker, whose experiences made the national news as the *Pascagoula Affair.* Pascagoula is situated close to a local major shipworks factory and Pensacola Naval Air Station in Florida, the kinds of facilities that have previously attracted the interest of surveilling UFOs. *See* **Hickson, Charles; Parker, Calvin**

Payntor, Bull

The pilot of Governor Ronald Reagan's plane, who confirmed the future president's sighting of a UFO in 1968. He said:

> I was the pilot of the plane when we saw the UFO. We were flying a Cessna Citation. It was maybe 9 or 10 o'clock at night. We were

> near Bakersfield, California, when Governor Reagan and the others called my attention to a big light flying a bit behind my plane.
>
> It appeared to be several hundred yards away. It was a fairly steady light until it began to elongate. Then the light took off. It went up at a 45-degree angle, at a high rate of speed. The UFO went from a normal cruise speed to a fantastic speed instantly. If you give an airplane power, it will accelerate, but not like a hot-rod, and that's what this was like. The object definitely wasn't another airplane. But we didn't file a report on the object because for a long time they considered you a nut if you saw a UFO.

See **Reagan, Ronald**

Pensacola, Florida

The naval air base near Gulf Breeze where the famous Gulf Breeze UFO incident took place and, some say, continued for the next few years well into the 1990s. *See* **Gulf Breeze Incident; Walters, Ed**

Pentagon

Central headquarters of the United States military general staff and Department of Defense. In 1947, it was the building where, according to Lt. Col. Philip Corso, a cache of debris from the crash of a UFO at Roswell was stored. On July 19 and 20, 1952, the Pentagon itself was the site of a UFO encounter when eight flying saucers flew over the building in a case that became known as the *Washington Overflights*. See **Corso, Philip J.; Department of Defense (DoD); Washington, D.C., UFO Overflights**

Pérez de Cuéllar, Javier

Secretary General of the United Nations and reportedly a witness to the "Brooklyn Bridge" abduction of Linda Cortile. *See* **Brooklyn Bridge Abduction Case**

Petukhov, A.

A Soviet commentator on UFOs from the All Union Council of Scientific and Technical Societies Commission, who in 1989 told the *Moscow: Mir:*

> UFOs have been seen to hover over ground objects, to chase or fly side by side with airplanes and cars, to follow geometrically regular trajectories, and to send out ordered flashes of light. In other words, such paranormals behave, from the viewpoint of human

beings, quite often showing capabilities yet beyond the reach of the machines built on the earth.

Philadelphia Experiment

USS Eldridge in drydock after she was sold to the Greek Navy.

A fictional time-travel story with a UFO connection that was perhaps based upon a marginally true story involving an experiment conducted at the Philadelphia Navy Yard in October of 1943. As the story goes, the Navy supposedly conducted an experiment attempting to render the destroyer escort *USS Eldridge* invisible. The experiment went tragically wrong when the ship and its crew were instead sent traveling through time to the 1980s. Though the ship eventually "returned" to 1943, crew members were killed or suffered deleterious after-effects from the trip.

Personnel at the Fourth Naval District believe that the questions surrounding the so-called "Philadelphia Experiment" arise from quite routine research, which occurred during World War II at the Philadelphia Naval Shipyard. Until recently, it was believed that the foundation for the apocryphal stories arose from degaussing experiments which have the effect of making a ship undetectable or "invisible" to magnetic mines. Another likely genesis of the bizarre stories about levitation, teleportation, and effects on human crew members might be attributed to experiments with the generating plant of a destroyer, the *USS Timmerman.* In the 1950s this ship was part of an experiment to test the effects of a small, high-frequency generator providing 1,000hz instead of the standard 400hz. The higher-frequency generator produced corona discharges and other well-known phenomena associated with high-frequency generators. None of the crew suffered effects from the experiment.

The source of the story, a man known as Carlos Allende, turned out to be a UFO hoaxer named Carl Allen, who had a proclivity for passing along his "knowledge" of UFOs, their means of propulsion, and the culture and ethos of the beings occupying the UFOs, all described in pseudoscientific and usually incoherent terms.

The fictional account of the Philadelphia Experiment, which became a best-selling book and two motion pictures, was enough to provoke the Office of Naval Research to release a statement to the press, in essence branding the whole affair "science fiction." *See* **Hoover, George W.; Jessup, Morris K.**

Phobos

One of the moons of Mars, which according to Soviet Col. Marina Popovich and in the opinion of Soviet scientists, is an artificial structure, and hollow as well.

Phobos II

The Soviet Phobos II probe seemed to have had a UFO encounter on its way to the Martian moon. *UFO Magazine* secured from respected Russian pilot Dr. Marina Popovich the very last reported photograph that the Russian probe took just prior to its apparent destruction. This photo showed what the Russians said was a "huge"

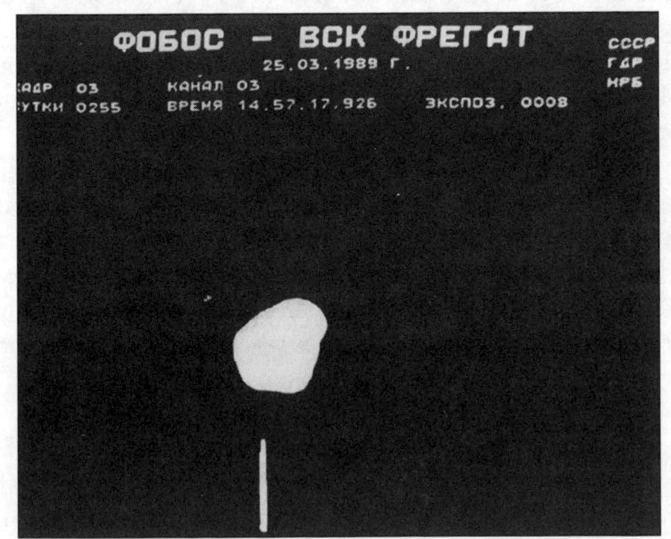

Phobos probe shows anomaly in this photo provided by Soviet pilot Marina Popovich.

fifteen-mile "UFO" that struck their probe, destroying it. The Russians as well as the Americans, many UFO researchers believe, are encountering difficulties with extraterrestrials who don't want human beings on Mars. *See* **Popovich, Marina**

Phoenix Lights

Phoenix Lights: an artist's interpretation.

These mysterious nighttime sightings over the greater Phoenix, Arizona, area in 1997 were captured on national television, mocked by then-Arizona Governor Fife Symington, and dismissed as a non-event by the military. However, the Phoenix Lights, which hovered menacingly over this desert city before finally fading out, were one of the most dramatic and important mass UFO events of recent times. The "Lights" achieved celebrity status because they came at a time when everyone seemed to own a home video camera, and everyone in Phoenix had videotape of the lights to show the media. Though their origin has never been fully explained, to many investigators they were UFOs, plain and simple. To others, they were flares dropped from military aircraft as part of an exercise.

According to researcher William F. Hamilton writing in *UFO Magazine*, sky watchers trying to catch a glimpse of comet Hale-Bopp instead noticed a "V" formation of white lights moving slowly over the Phoenix area from the northwest toward the southeast. Observers debated whether the lights defined the shape of a single, huge object blocking the background stars or whether they were individual objects. In either case, they covered an area large enough to be measured by entire city blocks.

The light formations were well within the flight-control radars at a number of local airports, and especially within radar contact of Luke Air Force Base outside Phoenix. Therefore, it was reasoned, there should be some record of radar contacts as well as the reports of visual sightings that were coming in from all areas around Phoenix. However, the day after the sightings, the Air Force denied it had recorded any extraordinary radar contacts on the night in question, and the public information representative at Luke Air Force Base announced that no planes had been scrambled to intercept any unidentified craft in the area—a denial which flew in the face of the testimony of witnesses who had observed jets trying to close with the lights.

A few months later a spokesperson for the Maryland Air National Guard announced that units from one of its air squadrons had indeed dropped high-intensity flares during military exercises they were conducting over a nearby test range. The flares were dropped in formation, the spokesperson said, suspended by tiny parachutes that would be invisible in the nighttime sky, thereby accounting for the anomalous lights observers had reported.

But, as Bill Hamilton and others have pointed out, flares don't ascend into the air, change direction, move horizontally, and then link up with one another. Moreover, photographic analysis revealed that the flare-drop locations were different from where the triangular lights had been seen. To this day, no satisfactory explanation of the Phoenix Lights has been provided by official sources. *See* **Barwood, Frances**

Pioneer 10

Launched on March 2, 1972 Pioneer 10, one of the most successful of NASA's space probes, was the first spacecraft to travel through the Asteroid belt and the first spacecraft to make direct observations and obtain close-up images of Jupiter. Famed through most of its mission as the most remote object ever made by man, Pioneer 10 is now over 7.6 billion miles away. If Pioneer 10 encounters intelligent life somewhere out there in deep space, it, like Voyager, will mark humankind's attempt to become a first contact with another species. Pioneer 10 made valuable scientific investigations in the outer regions of our solar system until the end of its science mission on March 31, 1997. *See* **National Aeronautics and Space Administration (NASA)**

Planet X

Also called by the Sumerian name *Nibiru,* Planet X has been described by Sumerian scholar Zecharia Sitchin as the "planet of the crossing," its definition in Sumerian. Planet X or Nibiru is the home of the Anunnaki, those entities the Sumerians referred to as their gods who delivered to them the collective knowledge of their culture. Anunnaki means "those who from heaven to earth came." According to Sitchin, Planet X has a 3,600 year orbit around the sun and has been the object of an ongoing search beyond Pluto. Sitchin has hypothesized that the gravitational effects of Planet X are observable on the orbits of Uranus and Neptune. *See* **Anunnaki; Sitchin, Zecharia**

Pleiadians

See **Alien Beings; Extraterrestrials; Grays; Meier, Eduard "Billy"**

Podkamennaya Tunguska River

On June 30, 1908, a large meteor, or so the explanation goes, exploded just above the earth's surface near this river in Tunguska, Siberia at approximately 7:40 a.m. The explosion was so great it was the equivalent of over a thousand times the explosion of the first atomic bomb over Hiroshima in World War II. Witnesses from five hundred miles away reportedly said that they could see a huge fireball explode over the horizon. Although scientists have said that it was a meteor, Japanese scientists believe that it was an alien spacecraft and have marked the site with a plaque commemorating it as an extraterrestrial event. *See* **Tunguska, Siberia**

Police Sightings of UFOs

Trying to obtain statements from official sources about UFO sightings is often an exercise in futility and obfuscation. Which is why, when law enforcement officers file a sighting report, the UFO community tends to sit up and take notice. The fact that policemen are all trained observers, sworn to high standards of conduct and responsibility, goes to the heart of their credibility. They are the one group that continues to experience the UFO phenomenon first-hand because they're the first ones called to the scene when a strange event takes place. *See* **Zamora, Lonnie**

Popovich, Marina

One of Russia's foremost test pilots and an expert on the Soviet space program and the discoveries that its scientists and cosmonauts made during the period before the Soviet Union's dissolution. At a 1990 press conference on the subject of Russian UFO sightings, Col. Popovich showed photographs of a cigar-shaped alien spacecraft that she said was fifteen miles long. The photos had been taken by the Soviet Phobos II space probe, which had mysteriously stopped func-

Marina Popovich holding a photo from Phobos II.

tioning soon after. In addition, Popovich has revealed that other Soviet satellites have taken photos of flying saucers, that Soviet scientists have concluded that flying saucers have been around for as long as our planet has existed, and that she, personally, has seen photographs of alien/human hybrid children. *See* **Phobos II; Russian Roswell**

Popovich, Pavel

A renowned Soviet cosmonaut. At the International MUFON UFO Symposium in 1992, Maj. Gen. Popovich said:

> Today it can be stated with a high degree of confidence that observed manifestations of UFOs are no longer confined to the modern picture of the world, or the simple refutation of the orthodox natural science paradigm. The historical evidence of the phenomenon, the singularity of its newly gained kinematic, energetic, and psychophysical features allows us to hypothesize that ever since mankind has been coexisting with this extraordinary substance, it has manifested a high level of intelligence and technology. The sightings have become the constant component of human activity.
>
> It's necessary to carry out the popular ufological enlightenment, since the probability for a meeting of a person with a UFO exists, and this person should be ready for this event. Precautionary measures are especially important. It's necessary to tell the truth, which has been distorted previously by the politically engaged sciences and most recently by ufological dilettantes.

Pounce

See **Projects, Government UFO**

Proctor, Loretta

William D. Proctor's mother. Loretta Proctor handled a piece of material from the Roswell crash debris and told UFO author and researcher Stanton Friedman that the material wouldn't burn and could not be cut with a knife. *See* **Roswell, New Mexico**

Proctor, William D.

Accompanied Mac Brazel on horseback to the Foster Ranch debris field. *See* **Roswell, New Mexico**

Projects, Government UFO

These are the various military or CIA study projects undertaken over the years and tasked to figure out the nature of the UFO phenomenon—whether these were craft from other worlds, what the intentions were of the aliens piloting these craft, and whether or not UFOs posed a threat to the national security of the United States. Some of the following projects are undocumented, but their existence has been pieced together from evidence that UFO researchers have been able to unearth over time:

Aquarius

Allegedly funded by the CIA in the 1950s to gather information about alien life forms through sightings and human encounters with aliens.

Bando

Assigned to gather medical information on alien races.

Blue Book

Begun in 1952, Blue Book was the descendant of previous projects Sign and Grudge, and was created to catalog and assemble information about UFO sightings. Over the years, it was learned that Blue Book was actually a filtering system; the more easily explained-away sightings were relegated to this office, while serious sightings and encounters that might have involved national security threats were never released to the public. In an effort to close down Blue Book, the Air Force commissioned the Condon Report, tasking Dr. Edward Condon to dismiss UFO sightings as being unworthy of scientific investigation and posing no threat to the U.S. When they received that report, the Air Force used it as a pretext to close down Blue Book, which they did in December 1969. The original Air Force Project Blue Book, for all of its shortcomings as a filter or a public relations enterprise, nevertheless brought a sense of organization to the mass of UFO sighting reports flooding all levels of government. Perhaps sometime in the future, the entire body of Blue Book data can be quantified to determine whatever common threads there may be, and analyzed according to some statistical standard.

Gleem

Created by President Eisenhower. Gleem later became Project Aquarius.

Grudge

On February 11, 1949, Project Sign was officially changed into Project Grudge in order to protect it from exposure to the media, to whom the existence of Project Sign was leaked. Grudge, also referred to in intelligence documents as *The Saucer Project,* was a continuation of the Air Force's attempt to study the UFO phenomenon. It was officially terminated on December 27, 1949, after the Air Force concluded, in Technical Report No. 102-AC-49/15-100, that flying saucers in and of themselves were not dangerous.

The Project Blue Book staff ina publicity photo from the 1960s.

Mogul

In a change of cover story, an Air Force report in 1994 named Mogul as the true source of the Roswell crash wreckage. Mogul was a series of top-secret unmanned balloon flights begun in the late 1940s, carrying sensors designed to monitor the Soviet Union's attempts to detonate a nuclear weapon. Moguls consisted of an array of balloons tethered together and stretching out, in some cases, over 500 feet. Although many of the scheduled Mogul launchings were canceled because of bad weather, those that were launched were ultimately recovered. Only one flight, Flight 9, still remains unaccounted for, and some researchers believe that it ultimately became the "Roswell balloon." Other researchers suggest that a Mogul could not have accounted for the amount and kind of debris recovered at Roswell.

Pounce

Evaluated alien space technology based upon debris recovered from crashed flying saucers.

Saucer

The military designation for Project Sign.

Sign

Referred to inside the military as *Project Saucer*, Project Sign was begun in February 1948 and lasted only one year before its existence was leaked to the press. To protect the classified nature of the group, the Air Force changed its name to "Grudge." The members of Project Sign were the first group inside the U.S. military assigned to study flying saucers. The project itself was begun as a result of a top-secret letter written by Lt. Gen. Nathan Twining, Chief of the Air Materiel Command, Air Technical Intelligence Center, at Wright-Patterson Air Force Base in Dayton, Ohio. In his letter, dated September 23, 1947, Twining wrote:

> The considered opinion of the Command concerning the so-called "flying discs" is that ... the phenomenon reported is something real and not visionary or fictitious ... there are objects probably approximating the shape of a disc, of such appreciable size as to appear as large as a man-made aircraft.

Sigma

Created for the purpose of establishing communications with extraterrestrial aliens. Supposedly a face-to-face meeting with aliens occurred in 1959 as a result of this project.

Snowbird

Established to test-fly alien spacecraft. Snowbird is supposedly still ongoing.

Propulsion Systems, Reverse-Engineered UFO

As a result of home-bred experiments, as well as the government's acquisition of propulsion systems from downed UFOs, the U.S. military has been experimenting with "exotic" propulsion systems since the early 1950s. Though conventionally powered aircraft still fill the skies and command the public's attention, government scientists and engineers have apparently achieved far more in the world of black projects. In fact, when former Lockheed president Ben Rich told a group of listeners that "we already have the technology to move among the stars," the group was shocked not so much by Rich's pronouncement as by the off-handed manner in which he said it. People claiming to have worked at Area 51 or been shown exotic propulsion technology have said that we have experimented with anti-gravity, nuclear fission, nuclear fusion, electromagnetic wave, and even cryogenic fuel systems, all with some promise.

One engineer, David Adair, who testified before Congress in 1997, said that he had been shown an "organic-sentient" propulsion unit which looked like the nervous system of a human being and was able to relate to the instructions from a pilot on a molecular level so as to read his or her intentions and translate them into direction, speed, and motion. *See* **Area 51; Rich, Ben**

Puthoff, Harold E.

Director of the Institute for Advanced Studies in Austin, Texas, Dr. Puthoff is considered one of the premier theoretical physicists in the field of vacuum zero point energy and has published several of the seminal papers in the area. A graduate of Stanford University, where he has been a research associate and lecturer, Puthoff also served several years as the director for the cognitive sciences program at SRI International. As a theoretical and experimental physicist he has worked in the areas of fundamental electrodynamics, quantum vacuum states, gravitation, cosmology, and high power microelectronics. He is co-author of the textbook *Fundamentals of Quantum Electronics* and is a Fellow of the Fetzer Institute. *See:* **Remote Viewing, Stanford Research Institute (SRI)**

Harold Putoff

Pye, Lloyd

Former novelist and screenwriter whose study of human origins, *Everything You Know Is Wrong, Book One: Human Origins,* posits a Sitchinesque view of the establishment of modern human life on earth rather than a Darwinist evolutionary theory. Pye believes, from his studies of hominids such as Big Foot and the Yeti, that modern humans are the result of the population of earth by beings from another planet rather than the product of indigenous evolution. Pye's most recent studies have concerned themselves with a skull specimen he has referred to as the *Star Child,* a skull discovered by a teenager in Mexico that displays characteristics so completely different from a normal human skull that Pye believes he has discovered a completely different species of hominid. The skull represents, Pye has said, either the skull of an alien creature or an alien-human hybrid, the result of ET interactions with humans that took place in primitive societies. *See* **Sitchin, Zecharia**

Pyramids

Many UFO researchers have found correlations between the pyramids of Egypt and what seem to be pyramids on Mars, the Moon, and Venus. Not surprisingly, from that has come the suggestion that perhaps the pyramid design antedates the Egyptians, who didn't invent the structure but apparently followed an already existing model.

Pyramids have been photographed on the lunar surface by American Lunar Orbiter probes which circled the moon in the 1960s. Richard Hoagland has said that the Face on Mars and its neighboring pyramids have geometrical correlations that match that of the Sphinx and the Egyptian pyramids. Also, according to Hoagland, in 1985 NASA discovered a complex of Sphinx and pyramids on the planet Venus, which also match the Egyptian models. This information was never released to the public, according to Hoagland.

In an article in *UFO Magazine* on the function of the Egyptian pyramids, engineer and author Christopher Dunn wrote that "the Great Pyramid is still the largest, most precisely built, and most accurately aligned building ever constructed in the world." Questions abound not only regarding the origin of the Great Pyramid but of the structure's purpose and the intent of its builders. Perhaps in the near future we will discover what it is about the pyramid form—found not only in Egypt but also in Mexico and Central America—that's so compelling that it's been duplicated elsewhere outside the Earth. *See* **Hoagland, Richard; Mars, Face on**

Quintanilla, Hector

Air Force major who became director of that service's Project Blue Book in 1963. Maj. Quintanilla claimed many UFOs were true unknowns, but was forced by circumstance to support the Air Force's stated position that they didn't exist. Quintanilla served as head of Blue Book until the project was shut down in 1969. *See* **Projects, Government UFO; Zamora, Lonnie**

Q, the Q Continuum

Although a fictional creation from the world of "Star Trek: The Next Generation" and "Voyager," Q as an extension of the continuum to which he belongs—as well as the character of Odo from "Deep Space Nine" and the Great Link of shape shifters—is a metaphor for the Jungian great link of the collective unconscious of all humanity, accessible by, according to those who practice the art, remote viewing or astral projection.

According to the philosophy of remote viewing, the power of psychic projection is an ability all human beings share, but it has been turned off because of our reliance upon language and hierarchical logic. Our ability to communicate with alien species is innate, many researchers have speculated, and, in fact, it can even be turned on by such an encounter. This ability, to join a great link of all sentient species, is what most governments fear human beings will learn from an encounter with alien races and in so doing obviate the need for the governmental and economic structures that control all civilized life on this planet. So, in order to protect those structures and those in power, an ongoing coverup of the existence of UFOs and our encounters with alien races persists, preventing human beings from joining the real great link or a similar version of "Star Trek's" Q continuum.

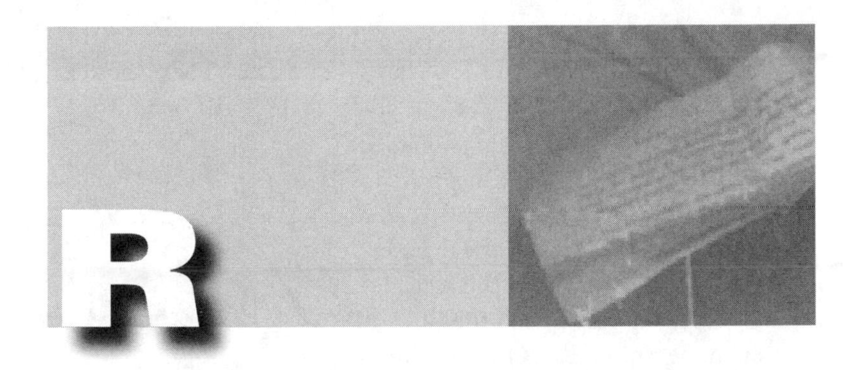

Rabochaya Tribuna

A leading Soviet newspaper cited in a U.S. Department of Defense memo concerning Col. Gen. Igor Maltsev's comments that UFOs piloted by intelligent beings "of some sort" had been visiting the Soviet Union. The DoD memo gives Maltsev's comments great credibility not only because they were reported in *Rabochaya Tribuna*, but because the general is the "Top Dog, the guy who runs the entire air defense system for Russia." *See* **Maltsev, Igor**

RAF Bentwaters

A NATO airbase on the North Sea coast of Great Britain where, in 1980, a bizarre and frightening series of events involving a possible encounter with alien beings from a UFO took place. *See* **Bentwaters Case**

Rael (Claude Vorilhon)

Founder of the UFO cult the Raelians, who espouse the belief that humanity was created from a cloning experiment involving hybridization between extraterrestrials and humans. Thus, the Raelians believe cloning is a way to reproduce and expand the abilities of the species.

Rael, *aka* Claude Vorilhon, was a professional race-car driver and sports journalist who claimed to have encountered small humanoid extraterrestrials near a volcanic crater in France. It was they who had created the earth's species, including "humans," and then "seeded" the

Rael, aka *Claude Vorilhon*

planet. This, they told him, was responsible for life on other planets as well—a version of the theory of panspermia. Vorilhon said that this revelation coincided with his own beliefs, inspiring him to spread the extraterrestrial gospel of the "Elohim," the name he gave the extraterrestrials; it comes from the Biblical Hebrew, "those who come from the sky."

Renaming himself *Rael,* Vorilhon founded a religious group claiming some 55,000 members worldwide and calling themselves "Raelians." Their belief system, as set forth by Rael, is based on the story of two Elohim mega-scientists, Yahweh and Satan, who got into a dispute over the value of their earthly creation. Yahweh believed the newly created species was essentially good, while Satan believed it was a genetic mistake. The Elohim chose him, Vorilhon said, to become the messenger of the "Good," like Jesus, because he was the hybrid offspring of a male Elohim and a female earthling.

Among Rael's preachings is the claim that the Elohim brought seven races from their own planet to inhabit earth, and now the human race and its descendants are about to become Elohim themselves. *See* **Clonaid; Crick, Francis**

Raelians

The cult founded by Claude ("Rael") Vorilhon after his self-described 1973 encounter with aliens. The Raelian doctrine is about the supremacy of science in all areas of human endeavor, except for love, which, Rael preaches, is a universal force for good.

The Raelians believe eternal life will come through cloning, a "gift" to humanity from the Elohim. Clonaid director Brigitte Boisselier, at first a skeptic about Rael's message and now his emissary in the world of science, says that behind the appearance of every new species there is evidence of some intelligent design. "These species aren't just appearing by chance," she proclaims. "Even Rael himself said that everything must be rational." *See* **Boisselier, Brigitte; Clonaid**

Ragsdale, James

Another self-proclaimed witness to the 1947 Roswell crash, who said he saw a UFO roar overhead on the night of July 4, 1947, while camping out in the desert with his girlfriend. Believing the object had crashed, Ragsdale and his girlfriend drove around and found it stuck in the side of a small hill. They thought it looked very unusual, and believed at first it might be an experimental military aircraft. In the dark desert night, they decided it best to leave and come back in the morning.

The next day they returned and could see the craft more clearly. It looked like an airplane with little crescent-shaped wings. The front end was collapsed, but the rear was intact. They also saw several bodies "about four feet long" which looked like "midgets." They both picked up pieces of debris, and Ragsdale described the material as unlike anything he had ever seen. "You could take that stuff and wad it up and it would straighten itself out," he said.

Shortly thereafter the military showed up in force. Ragsdale said he was afraid that he and the girl would get into trouble, so they threw down the pieces of wreckage and hid, eventually watching the military scour the area clean of debris and remove the bodies. Ragsdale's story has since been attested to by family members. *See* **Roswell, New Mexico**

Rak, Chuck

One of the abductees in the "Allagash Affair," who claimed he was taken aboard a spacecraft while canoeing across a lake on the Allagash Waterway in Massachusetts. The full account of his story appears in Raymond Fowler's book *The Allagash Abductions*. *See* **Allagash Affair**

Ramey, Roger

Commander of the Eighth Army Air Force headquartered at Fort Worth, Texas, who ordered the Roswell debris transferred to his command and then sent it on to Wright-Patterson Air Base in Ohio. After Gen. Clements McMullen passed orders along to Ramey "to get the press off our backs," Gen. Ramey then ordered base meteorologist Irving Newton to pose with substituted debris and tell the press that the object recovered from Roswell was really a downed weather balloon. *See* **DuBose, Thomas J.; Johnson, J. Bond; Newton, Irving; Roswell, New Mexico**

Ramey Telegram

The justifiably now-famous July 8, 1947, photograph taken by J. Bond Johnson of Brig. Gen. Roger Ramey and Col. Thomas J. DuBose examining debris from the Roswell crash. In Ramey's hand is a Western Union telegram, which, when the photo was enlarged half a century later, certain words are obvious: ("..." indicates type is indecipherable):

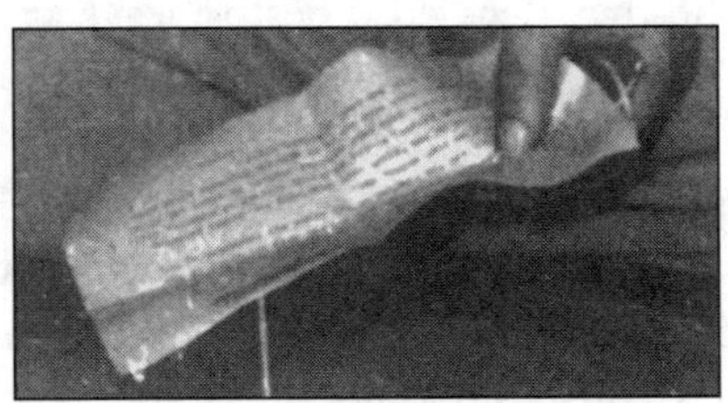

A small portion of the famous Roswell "balloon debris" photograph from 1947 showing a partially obscured telegram.

> ... 4 HRS THE VICTIMS OF THE ...
> YOU FORWARDED TO THE ... AT
> FORTWORTH, TEX.
> ... THE "CRASH" "STORY" ... FOR 0984
> ACKNOWLEDGES ... EMERGENCY
> POWERS ARE NEEDED SITE TWO
> SW MAGDALENA, N MEX.
> ... SAFE TALK ... FOR MEANING OF
> STORY AND MISSION ... WEATHER
> BALLOONS SENT ON THE ... AND
> LAND ... ROVER CREWS.

This is the order commanding Brig. Gen. Ramey to come up with a weather balloon cover story that would effectively bury the previously released information on the Roswell "crashed disc." *See* **Johnson, J. Bond**

Randle, Kevin

UFO author and researcher who personally interviewed many of the eyewitnesses to the Roswell crash. Author of *The Truth About the UFO Crash at Roswell*, among many others.

Ratsch, Donald B.

UFO investigator and researcher whose specialty is collecting film footage from NASA space missions which show UFOs flying nearby. *See* **Fastwalkers; NASA**

Reagan, Ronald

The fortieth president of the United States and former governor of California, Ronald Reagan is the only president besides Jimmy Carter to admit publicly that he'd seen a flying saucer. Reagan was the only president to hypothesize publicly, albeit metaphorically, about the possibilities of hostile extraterrestrials posing a threat to earth.

In a speech before the United Nations General Assembly on September 21, 1987, Reagan used the example of an alien attack on humanity as a rationale for governments to put aside their differences and come together for the common good. He did so with the full knowledge that many of the assembled press would take his remarks derisively. Nevertheless, Reagan also knew that most, if not all, of the assembled diplomats represented governments that had their own issues in dealing with the UFO "threat." Reagan had used the same analogy before during a speech at his summit meeting with Soviet president Mikhail Gorbachev in Geneva in 1985.

Unlike his predecessors, some of whom might have been more inclined to disclose the existence of UFOs to the American people, Ronald Reagan proposed the development of a planetary defense system called the Strategic Defense Initiative (SDI), or, as it was dubbed by the media, "Star Wars." Though the program was touted as a "shield" against incoming Soviet ballistic missiles, many UFO researchers thought that it was public cover for its real mission: a defense against hostile UFOs. In fact, the Soviets were actually inclined to support the project, despite their public protestations, once Reagan offered to make it available to them.

In 1972, then-Governor Ronald Reagan of California described his UFO sighting to Norman C. Miller, the Washington Bureau Chief of the *Wall Street Journal*:

> I was in a plane last week when I looked out the window and saw this bright light. It was zigzagging around. I went up to the pilot and asked him if he had ever seen anything like that. He was shocked and said he hadn't. So I told him to follow it, which we did for several minutes. It was a bright white light. We followed it

to Bakersfield, and all of a sudden to our utter amazement, it went straight up into the heavens. When I got off the plane I told Nancy about it.

Regehr, Ron

An aerospace engineer who has worked inside the defense industry for over twenty-five years, Regehr was a MUFON investigator who examined the 1947 J. Bond Johnson photograph of the telegram held by Gen. Roger Ramey and hired a team of experts to interpret what the telegram said from an enlargement of the photo.

As an engineer, Regehr analyzed performance reports on the Air Force's Defense Support Program satellites. He learned that the satellites had repeatedly picked up objects coming in from deep space, following trajectories totally unlike that of meteors. Regehr said that these objects frequently slowed down, changed direction, and then left earth's atmosphere at less than escape velocity. The reports confirmed for Regehr the existence of what Jacques Vallee has called *fastwalkers*. *See* **Ramey Telegram; Fastwalkers**

Remote Viewing

The controlled process of leaving one's body (out-of-body projection) and traveling to other places and times to observe what's there. According to those who have practiced this skill, one can discern what others are thinking and feeling and understand their thoughts and motivations while in this mode.

International Remote Viewing Association Founders Rear left to right: Hal Puthoff, David Hathcock, John Alexander, Lyn Buchanan, Paul H. Smith, Skip Atwater, Angela Thompson-Smith, Marcello Truzzi. Front, seated, left toright, Russell Targ, Stephen Schwartz.

Although remote viewing is probably the same process as the Native Americans' "spirit-walking," the releasing of the spirit from its physical body so it can roam to different places, it became weaponized in the 1970s by both the United States and the Soviets, possibly as a result of the discovery of the extraterrestrials' similar abilities.

Early remote viewing pioneers include Ingo Swann, Russell Targ, Hal Puthoff, Joe McMoneagle, and Gen. Albert N. Stubblebine. Governments, aware that enemy remote viewers might be spying on their facilities, have sometimes come up with innovative defenses to protect against intelligence gathering. It's been reported that the Soviets used to camouflage their military facilities with balloons featuring cartoon characters so that when American remote viewers paid a "visit," they would believe they'd accidentally come upon an amusement park and not a top-secret location. *See* **Native Americans; Puthoff, Hal; Swann, Ingo**

Rendlesham Forest, England

The heavily wooded area adjacent to the NATO air base at RAF Bentwaters where, allegedly, a UFO appeared on December 26, 1980, facilitating contact with the base command. *See* **Bentwaters Case**

Renglin, Liang

A professor at Guangzhou Jinan University in China who, in the *China Daily* on August 27, 1985, said:

> More than six hundred UFO reports have been made in China during the past five years. UFOs are an unresolved mystery with profound influence in the world.

See **China, UFOs in**

Report Number 14

A U.S. Air Force internal report that refutes the service's public stance regarding UFOs: that nearly all UFO sightings are of naturally occurring phenomena. The data set forth in Report Number 14 demonstrates that, according to the Air Force's own figures, psychological explanations for UFO experiences and outright hoaxes account for only 1.2 percent of the reported sightings, while "Insufficient Information" and "Other" account for over 40 percent. By the report's own data, then, more than 50 percent of reported UFO sightings cannot be exlained or identified. This also contradicts the Air Force's stated rationale for closing down Project Blue Book and using the findings of the Condon Report to provide them with the pretext for doing so. *See* **Condon Report; Projects, Government UFO**

Reptilians

Extraterrestrial beings, as reported by contactees, resembling six-foot-tall lizard-like creatures, who have visited earth and heavily contributed to the DNA makeup of human beings. Contactees say the Reptilians are a race of explorers and seeders of life who have been in contact with the governments of earth, and have chosen not to reveal their existence to humanity. Controversial lecturer and author David Icke believes that many governments are, in fact, secretly manipulated by these alien entities. *See* **Alien Beings; Icke, David**

Rich, Ben

Former head of Lockheed and author of the book *Skunk Works*, which told the story of how our most technologically advanced aircraft were developed in secret. Rich also claimed that American technology is so far advanced that our propulsion systems could "send ET home," but these advances are being kept secret from the public. *See* **Skunk Works**

Rivers, Mendel

Chairman of the House Armed Services Committee. In 1966, Rivers received a letter from then-Representative Gerald Ford of Michigan recommending that a House committee be established to investigate the UFO phenomenon.

Robertson Panel

A panel convened in 1953 under the chairmanship of physicist H. P. Robertson, chairman of the U.S. Office of Scientific Intelligence, to study the veracity of UFO reports and any potential threats UFOs might pose to the national security of the United States. Members of the panel included Lloyd Berkener, a reported member of MJ-12, and astronomer J. Allen Hynek.

The panel interviewed military officers as well as scientists, and reviewed photographs of reported UFOs, written reports of sightings, and technical data associated with the UFO phenomenon. It concluded that UFOs could be explained away as natural phenomena, and that there was no credible evidence of an extraterrestrial presence on earth or an alien threat to humanity.

However, the panel did find that UFO organizations themselves posed something of a threat because they were capable of being manipulated by enemies of the United States into inducing mass hysteria about an alien threat. It was for this reason the panel suggested that all UFO reports should be debunked.

Many UFO researchers believe that the presence of a member of MJ-12 among the panel members was an example of the government's attempt to manipulate and then shut off any disclosure of UFO information and use formal scientific panels as a method of debunking legitimate scientific inquiries. *See* **Berkener, Lloyd V.; Central Intelligence Agency, History of UFOs in; Hynek, J. Allen; Majestic**

Rodden, Jack

A photographer in Roswell, whose father, also a photographer, did occasional work for the U.S. Army. It is thought that the elder Rodden took pictures of the Roswell crash site for the Army, photographs that proved the existence of an alien spacecraft and of the extraterrestrials who piloted it.

Roosevelt, Franklin Delano

Thirty-second president of the United States and former governor of New York. During the 1942 flying-saucer sightings over Los Angeles, Gen. George C. Marshall memoed President Roosevelt to tell him that the objects overhead were moving at over 200 miles per hour, a speed that witnesses said was nowhere near the actual speed of the objects. Whether FDR knew or believed that these objects were flying saucers and not Japanese bombers is unknown. However, FDR was a believer in the paranormal and, at the beginning of his final term in office, consulted with a medium to find out how long he had to live.

Roswell, New Mexico

A thriving little city near the White Sands Proving Grounds and Alamogordo, where the first atomic bomb was detonated, Roswell has been associated with the July 1947 crash of a UFO in the desert nearby, referred to as the *Roswell Incident.* Whether it was a flying saucer that crashed, a secret military test plane, a standard weather balloon, or a Mogul surveillance balloon, the Roswell Incident has secured its place in American folklore and is regarded as perhaps the place of humanity's first contact with technology and beings from another world and the impetus for an all-encompassing government cover-up.

There are many different versions of the Roswell crash story. For the purposes of bringing together as many versions as possible, this is the timeline of the story as described by Harold Burt in his book *Flying Saucers 101*.

Tuesday, July 1, 1947

Military installations at White Sands, Alamogordo, and Roswell, New Mexico, track high-speed objects on radar, all flying at speeds well above what current-day jet fighters are

Arroyo into which the flying saucer allegedly crashed outside of Roswell

capable of doing. With the objects penetrating highly secure air space at will, all three facilities go on full alert.

Wednesday, July 2, 1947

Civilians spot disk-shaped craft flying over Roswell. Numerous reports come in from the public. The number of radar targets increase; there is now almost a constant presence of the unidentified objects showing up on radar.

Friday, July 4, 1947

Late at night during a tremendous thunderstorm, a fast-moving object disappears from military radar screens. Local citizens hear a loud, unusual explosion and see an aerial object crash into the desert. Radar observer and former Army CIC Sgt. Frank Kaufman watches a blip on his radar screen pulsate, enlarge, and then disappear.

Col. William Blanchard believes an enemy aircraft has penetrated his defense systems and has crashed nearby. He immediately sends out a team to investigate. Blanchard desperately wants to know what kind of aircraft can make hairpin turns at 3,000 miles per hour, which he has just observed on radar.

SATURDAY, JULY 5, 1947

About 4:00 A.M., the military team reaches the still-smoldering wreckage. They call local Chaves County Sheriff George Wilcox and tell him to send out the fire department.

Site claiming to be official UFO Crash site in Roswell

One of the military team members is Frank Kaufman. At the site he observes an almost completely intact object partially buried in the sand as the result of a terrific impact. The craft appears to be crescent-shaped or a rounded-delta shape.

There are several small, dark-gray bodies sprawled on the sand near the craft. Another one lies near a breach in the hull. One being is still alive and tries to flee. The panicked soldiers fire their weapons, killing the alien.

(Although the preceding is based on Frank Kaufman's testimony, his story has been largely discredited by Roswell researchers. However, portions of it do comport with many other eyewitness claims regarding the discovery of a second Roswell crash site, this one containing the largest piece of wreckage—the saucer—and the recovery of its alien passengers by the military.)

SATURDAY, JULY 5, 1947

Early in the morning, William (Mac) Brazel rides out on horseback across the Foster Ranch to check on his livestock following the previous night's thunderstorm. He's accompanied by seven-year-old William D. Proctor. They discover a large amount of strange debris strewn across the range, debris that the sheep refuse to walk through. Brazel ends up having to herd the ranch's sheep almost a mile around the site to get them to water.

The debris consists of pieces of lightweight material that resembles tinfoil, and lightweight I-beams with unknown symbols engraved in their sides. Brazel gathers up some to take home and also makes up a box to take to Sheriff George Wilcox in Roswell.

SUNDAY, JULY 6, 1947

Sheriff Wilcox thinks the debris is strange enough that he calls the nearby Roswell Army Air Base and speaks to Maj. Jesse Marcel, the base's intelligence officer. Marcel relays the information to his commander, Col. William Blanchard. Blanchard orders Marcel to accompany Brazel back to the ranch to take a look around. Marcel is joined by Capt. Sheridan Cavitt, an officer with the Counter-Intelligence Corps. The men drive separate cars out to the site.

They arrive at the Foster Ranch after dark and spend the night in a small cottage before venturing out to the debris field the following morning.

Meanwhile, Col. Blanchard receives an order from Maj. Gen. Clements McMullen, Commander of the Strategic Air Command, to retrieve some of the debris from Sheriff Wilcox's possession and send it immediately to Col. Thomas DuBose at the Fort Worth Army Air Field. Wilcox and his

Interviews about the crash are often filmed at one of the alledged sites.

A party atmosphere descends on the small town of Roswell every July to commemorate and discuss the events that occurred in 1947.

family are told in no uncertain terms to remain silent about the events of the previous few days. The debris is sealed in a special lead-lined courier pouch and flown to Texas.

At Fort Worth Army Air Field, Col. Alan D. Clarke, the base commander, personally meets the plane carrying the Roswell crash debris. He takes charge of it, hops aboard a B-26 and flies it to Washington, D.C., where he's met by Maj. Gen. McMullen.

MONDAY, JULY 7, 1947

Early in the morning, Mac Brazel takes Marcel and Cavitt to the site of the debris field. The wreckage is strewn over an area covering at least a square mile. The officers fill their vehicles with debris and head back to the base. Maj. Marcel is impressed enough by the recovered material to make a side trip to his house, arriving there at 2:00 A.M. on Tuesday morning.

Marcel awakens his wife and 11-year-old son, Jesse, Jr., to show them the material. It is so strange that, years later, Marcel, and then his son, would say they believed it was not from this planet.

TUESDAY, JULY 8, 1947

Marcel links up with Cavitt and the two officers deliver their respective car-loads of debris to the air base around 6:00 A.M. They brief Col. Blanchard, and Blanchard issues an order cordoning off the debris field with armed guards. An air search is also begun to see if more debris can be found.

Another view of famed RAAF Hangar 54 as it looks today.

At approximately the same time, Sheriff Wilcox orders two of his deputies out to the debris field, where they find military sentries posted who turn them away. The deputies observe a large circular imprint in the ground where something has impacted and which has turned the surrounding sand into glass. At the debris field, Army personnel continue to fill wheelbarrows full of wreckage and load it onto trucks.

At around 9:00 A.M. Col. Blanchard holds a staff meeting. He orders Lt. Walter Haut, the Roswell Army Air Base public information officer, to issue a press release saying the Army has recovered the remains of a crashed flying disk. Haut delivers the news release to local radio stations at approximately 11:00 A.M., which sets off a firestorm of headlines and stories in the late and evening newspaper editions.

Later that same afternoon, after the recovered debris has reached Fort Worth and the office of Gen. Roger Ramey, J. Bond Johnson, a young reporter for the *Fort-Worth Star Telegram*, is dispatched to Gen. Ramey's office to take pictures. He sees the general's floor covered with debris of some kind, which fills the room with a strong, burnt odor. Using a Speed-Graphic camera, Johnson takes a total of eight photographs on the two-sided film plates.

(Years later, Johnson said, although it has not been substantiated, that he believes the photos he took that day were of the real saucer debris. If a switch to balloon debris had been made for subsequent photos, it occurred after he left. Col. DuBose, in an interview in the November 24, 1991 edition of *Florida Today*, said that the real Roswell debris had already been transferred to Washington, D.C., by the time the weather balloon debris was displayed for the press.)

TUESDAY, JULY 8, 1947; 3:00–5:00 P.M.

Gen. Ramey announces that "the flying disc has been sent to Wright-Patterson Air Base in Ohio." Gen. McMullen calls Col. DuBose from Washington and orders him to tell Ramey to send some of the debris to Washington immediately and to make up a story "to get the press off our backs." At around 5:00 P.M., the Pentagon phones Roswell Army

Gen. Roger Ramey and Maj. Jesse Marcel pose for a photo over the debris. In Gen. Ramey's left hand is the telegram he allegedly received from his superiors in Washington ordering the coverup of the flying saucer crash.

Air Base and orders Col. Blanchard to halt all press releases regarding the crashed saucer.

Maj. Marcel learns that he will have to transport the full cache of debris to Fort Worth Army Air Field, using a specially-prepared B-29 bomber. The carloads of debris will ultimately fill up half of the B-29. By 3:00 P.M., Marcel has arrived at Fort Worth with the debris, delivered it, posed for photographs, and left.

At about 5:30 P.M., Gen. Ramey summons a young weather officer, Irving Newton, to his office to pose for new photographs that comport with the Army's cover story, now in place, that the Roswell "disc" is in fact a weather balloon. Sometime before Newton arrives, the remains of a standard weather balloon are spread out on Ramey's floor and a new set of photos is taken. At 8:00 P.M., Ramey calls a press conference to announce that a mistake has been made and asks Newton to identify the object as a balloon for the reporters who are gathered there. None of the reporters challenges him on the fact that it has taken over five hours to catch the "mistake." At 10:00 P.M., ABC News declares that the Roswell wreckage is a weather balloon. And as far as the public is concerned, the story stops right there.

Twenty years later, retired Army Air Force major Jesse Marcel decides to go public. He tells his story to Stanton Friedman, thus beginning a new chapter in the Roswell story. *See* **Roswell Crash Press Release**

Roswell Army Air Field

Home of the 509th bomber squadron, at the time the world's only active nuclear bomber group. Roswell Air Field was the location where the Roswell crash debris was first taken, and from where it was then sent to Fort Worth and finally to Wright-Patterson.

Roswell Crash Press Release

The text of Lt. Walter Haut's press release, dated July 8, 1947:

The many rumors regarding the flying disc became a reality yesterday when the Intelligence office of the 509th Bomb Group of the Eighth Air Force, Roswell Army Air Field, was fortunate enough to gain possession of a disc through the cooperation of one of the local ranchers and the sheriff's office of Chaves County.

The flying object landed on a ranch near Roswell sometime last week. Not having phone facilities, the rancher stored the disc until such time as he was able to contact the sheriff's office, who in turn notified Maj. Jesse A. Marcel of the 509th Bomb Group Intelligence Office.

Action was immediately taken and the disc was picked up at the rancher's home. It was inspected at Roswell Army Air Field and subsequently loaned by Maj. Marcel to higher headquarters.

Roswell Spacecraft

Although first reports described it as a "flying disc," the Roswell craft was actually more of a crescent shape. A few years following the crash, Jack Northrop built his YB-49 bomber, which resembled illustrations of what the Roswell craft looked like. Today, the bat-winged B-2 bomber and F-117 Stealth fighter both resemble the craft that was reported to have crashed at Roswell. *See* **B-2 Spirit Bomber; Northrop, Jack**

Rowe, Frankie

Roswell crash witness Frankie Rowe's father was a firefighter who arrived at the crash site just before the military retrieval team did, and saw alien bodies. Frankie Rowe also handled some of the alien fabric that the firefighters brought back to the firehouse after they returned. She said that she and her family were threatened with death by a man wearing a military uniform. *See* **Children of Roswell**

Ruby, Doug

Retired Air Force and commercial airline pilot Ruby believed that crop circles were schematic diagrams that actually held the key to designing spaceships. Instead of looking at them as two-dimensional blueprints, he believed the secret lay in seeing them as three-dimensional objects. Ruby also believed that each crop circle was only a component of a larger design.

By combining photos of separate crop circles to see how they might fit together, and by taking them as three-dimensional schematics, he was able to build a scale model of a spaceship that looked remarkably like the "beamships" that Billy Meier had described. Ruby's diagrams also suggested the way the craft propelled themselves, using counter-rotating rings around their saucer edges to generate electromagnetic waves. That would also account for the numerous reports of flying saucers interfering with electrical devices. *See* **Beamships; Crop Circles**

Ruppelt, Edward J.

U.S. Air Force project coordinator for flying saucer investigations, head of Project Blue Book, and author *The Report on Unidentified Flying Objects*. In his book, Capt. Ruppelt claimed that the government knew about the existence of flying saucers and covered up the fact. After Ruppelt began working for a company doing contract work for the Air Force, his attitude towards UFOs changed, perhaps under pressure from the Air Force, and he retracted many of his earlier assertions.

Russian Consulate, San Francisco

Site of a historic 1990 press conference on the subject of UFOs held by Col. Marina Lavrentevna Popovich during which she showed photographs of cigar-shaped alien craft taken in space. The photos were taken by a Russian space probe that disappeared shortly after taking the pictures. She also said that Russian scientists had concluded that UFOs had been around for as long as our planet's existence and that she herself had seen photographs of alien/human hybrid children. *See* **Popovich, Marina**

Russian Roswell

In late 1991 after word had leaked out that what seemed have been a UFO crashed in the Tien Shan mountains, two expeditions set out to find the object. The first expedition actually reached the site, but because of intense radiation, they could not approach the object and vacated the area. Another expedition tried to reach the craft about a year later, but when they reached the area they discovered that the object had been removed.

The story of this crash and the nature of the craft that managed to elude Russian MiG-29s in a high-speed chase across the Caspian Sea only to crash and be recovered by the military corresponds to the crash at Roswell, New Mexico, in 1947. Both are shrouded in mystery. *See* **Hesemann, Michael; Popovich, Marina**

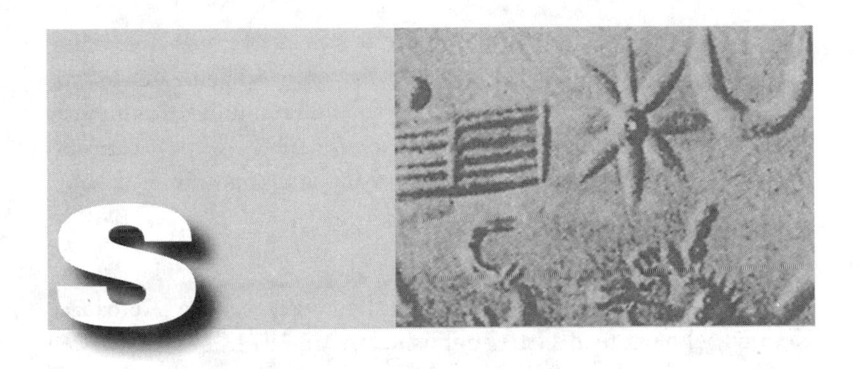

S-4

Sheltered securely within the U.S. Air Force's top-secret Area 51 in Nevada, site S-4 is said to house laboratories that are responsible for reverse-engineered downed alien spacecraft. According to some reports, the U.S. government has extraterrestrials at the facility assisting military scientists working there. S-4 is also said to contain camouflaged hangars built into a nearby mountainside holding the remains of several different kinds of alien saucers, one of which was studied by engineer Robert Lazar. *See* **Area 51; Lazar, Robert**

Sagan, Carl

Astronomer, author, and one of America's most prolific communicators of complex scientific theories to the general public. Sagan's "Cosmos" television series on PBS is still considered a landmark work, while his novel *Contact* offered a mainstream scientist's measured view of what mankind's first contact with extraterrestrials might be like.

Though Sagan played the role of UFO debunker to the hilt, many researchers believe that he was closely tied to those people in the know about UFOs, and that his denials of the existence of UFOs often rang hollow if not downright false. Was Sagan privy to secrets he could not discuss? Was silence the price he paid for the opportunity to consult on an alien presence on planet Earth? A clue may lie in a statement Sagan once made on "Cosmos" in 1980:

I believe the search for extraterrestrial intelligence to be an exceedingly important one for both science and society ... But I do not believe that the most efficient method of examining this topic is via the UFO problem. The best hope for such investigations is NASA's unmanned planetary program and attempts at interstellar radio communications.

Salutun, Commander J.

Indonesian Air Commander, the Secretary of the Aerospace Council, who wrote in a letter published in ˆ in 1974, "I am convinced that we must study the UFO problem seriously for reasons of sociology, technology, and security."

Salyut-6 Space Station

An early generation of the Soviet's manned space lab. On May 5, 1981 cosmonaut Vladimir Kovalyonok observed a melon-shaped UFO following the station in orbit, and watched as it changed shape, then vanished in what appeared to be an explosion. *See* **Kovalyonok, Vladimir; Russian Roswell**

Samford, John

After the UFO overflights of Washington, D.C., in 1952 and the military's apparent inability to protect the Capitol air space from intrusions by technologically superior aircraft, Director of Air Force Intelligence Gen. John Samford held a press conference during which he tried to explain that the objects were, in fact, only optical illusions, describing the phenomenon as a temperature inversion—despite the fact that the flying objects had been confirmed as real targets by five different radar stations, as well as by the military pilots who had been scrambled to intercept them. *See* **Washington, D.C., UFO Overflights**

San Augustin, New Mexico

One of the four sites in New Mexico where, it is alleged, a flying saucer crashed in 1947. According to some researchers the crash at San Augustin occurred several days before the Roswell crash, resulting in the military's being called to the site and then "sanitizing" it. The veracity of a crash at San Augustin can be established by the fact that the witness who first talked about what he saw there was U.S. Conservation Corps soil engineer Grady Barnett, who had arrived at the scene shortly after the crash and saw a large metallic object stuck in the ground, along with several dead bodies with oddly shaped heads and no hair. *See* **Barnett, Grady**

Sandia National Laboratory

One of the laboratories that has analyzed so-called alien implants. *See* **Implants, Alien; Leir, Roger**

San Luis Valley, Colorado

Site of the first modern cattle mutilation case in 1967 and a UFO-paranormal "hot spot." The San Luis Valley lies adjacent to NORAD (North American Air Defense Command) headquarters, as well as sensitive military facilities in both Colorado and Utah, some of which have been touted as being the "new Area 51," since it's believed the original Nevada site has transferred most of its operations as a result of increasing publicity. *See* **Area 51; Cattle Mutilations; Howe, Linda Moulton**

Santa Claus

Codename for UFOs, first used by Mercury astronaut Wally Schirra to indicate the presence of a flying saucer near the space capsule. Astronaut James Lovell, aboard the historic Apollo 8 mission in December 1968, also reported sighting "Santa Claus."

Santilli, Ray

British television producer who first distributed the "Alien Autopsy" film. Santilli claims that in 1995 he was approached by a man who had fifteen reels of 16-mm film depicting the actual 1947 autopsy of a Roswell extraterrestrial. The film, while compelling, has been branded as a hoax by many UFO researchers. *See* **Alien Autopsy Film**

Ray Santilli holds a film canister from his highly disputed project.

Santorini, Paul

A nuclear physicist and engineer who also investigated UFOs spotted over Greece in 1947. Dr. Santorini was quoted by author Raymond Fowler in *UFO's: Interplanetary Visitors* as having said:

We soon established that they were not missiles. But before we could do any more, the Army, after conferring with foreign officials, ordered the investigation stopped. Foreign scientists flew to Greece for secret talks with me. A world blanket of secrecy surrounded the UFO question because the authorities were unwilling to admit the existence of a force against which we had no possibility of defense.

Sarbacher, Robert

Physicist, president, and chairman of the board of the Washington Institute of Technology, dean of the graduate school at Georgia Institute of Technology, and a consultant to U.S. Army Research and Development in the 1950s and Lieutenant General Arthur Trudeau in the 1960s.

Dr. Sarbacher was reportedly connected to MJ-12, the group of high-level researchers and scientists charged with examining the UFO "threat" in 1947. Sarbacher's September 15, 1950 interview with Canadian engineer Wilbert Smith confirmed for many researchers the existence of UFOs.

The interview occurred long before any public discussions about the crash at Roswell, at a time when the U.S. government was desperately trying to suppress any stories about UFOs, especially stories from official government sources.

In a letter to UFO researcher William Steinman in November 1983, Dr. Sarbacher described the level of his participation in UFO-related events:

> I had no association with any of the people involved in the recovery and have no knowledge regarding the dates of the recoveries.
>
> Regarding verification that persons you list were involved, I can say only this: John von Neumann was definitely involved. Dr. Vannevar Bush was definitely involved, and I think Dr. Robert Oppenheimer also.
>
> Although I had been invited to participate in several discussions associated with the reported recoveries, I could not personally attend the meetings.
>
> About the only thing I remember at this time is that certain materials reported to have come from flying saucer crashes were extremely light and very tough. I am sure our laboratories analyzed them very carefully. There were reports that instruments or people

> operating these machines were also of very light weight, sufficient to withstand the tremendous deceleration and acceleration associated with their machinery. I remember in talking with some of the people at the office that I got the impression these "aliens" were constructed like certain insects we have observed on earth, wherein because of the low mass the inertial forces involved in operation of these instruments would be quite low.
>
> I still do not know why the high order of classification has been given and why the denial of the existence of these devices.

Wilbert Smith's interview with Sarbacher, and Smith's memo to the Canadian government concerning the existence of UFOs are crucial documents in UFO research. Additional material from the Smith-Sarbacher exchange follows:

> My association with the Research and Development Board under Dr. Compton during the Eisenhower administration was rather limited so that although I had been invited to participate in several discussions associated with the reported recoveries, I could not personally attend the meetings. I am sure that they would have asked Dr. von Braun, and the others that you listed were probably asked and may or may not have attended. This is all I know for sure.

NASA's Viking space probe that was test flown at Roswell in 1972.

I did receive some official reports when I was in my office at the Pentagon, but all of these were left there as at the time we were never supposed to take them out of the office. I do not recall receiving any photographs such as you request so I am not in a position to answer.

I recall the interview with Dr. Brenner of the Canadian Embassy. I think the answers I gave him were the ones you listed. Naturally, I was more familiar with the subject matter under discussion, at that time. Actually, I would have been able to give more specific answers had I attended the meetings concerning the subject. You must understand that I took this assignment as a private contribution. We were called "dollar-a-year men." My first responsibility was the maintenance of my own business activity so that my participation was limited.

Saucer, Project

The defense intelligence term for Project Sign, which began in February 1948 and lasted a year before its existence was leaked to the press. The name was then changed to Project Grudge, even though the military still kept referring to it, even during Grudge, as *the Saucer Project*. *See* **Projects, Government UFO**

Schirra, Wally

During the early Mercury missions, astronaut Wally Schirra was the first astronaut to use the term *Santa Claus* to refer to a UFO because NASA had forbidden the astronauts to use the term *UFO* in radio transmissions. *See* **Apollo Program; Lovell, James**

Schmitt, Donald

Author and researcher, who, along with Kevin Randle, interviewed many of the original Roswell witnesses for their book *The Truth About the UFO Crash at Roswell.*

Schreiber, Bessie Brazel

Mac Brazel's daughter, who told UFO researcher and author Stanton Friedman, "Some of these pieces [of Roswell saucer debris] had something like numbers and lettering on them, but there were not words we were able to make out. The figures were written out like you would write numbers in columns, but they didn't look like numbers we use at all." *See* **Children of Roswell**

Schuessler, John F.

John Schuessler is one of the founding members of the Mutual UFO Network (MUFON). Schuessler has been involved in the U.S. manned space program since 1962 and has been a very active UFO researcher since 1965.

John Schuessler, current head of the Mutual UFO Network.

Schuessler is a member of the National Institute for Discovery Science (NIDS) Science Advisory Board and is currently the deputy director for administration, a consultant in astronautics, and a member of the MUFON board of directors. *See* **Mutual UFO Network (MUFON); National Institute for Discovery Science (NIDS)**

Schulgen, George

In an official memo about the origin of UFOs written on October 28, 1947, Gen. Schlugen of the Air Intelligence Requirements Division wrote:

> While there remains a possibility of Russian manufacture, based on the prospective thinking and actual accomplishments of the Germans, it is the considered opinion of some elements that the object may in fact represent an interplanetary craft of some kind.

Schultz, C. Bertrand

A paleontologist working in the Roswell area on Saturday, July 5, 1947, who recorded a conversation in his diary with Texas Tech archaeologist Dr. W. Curry Holden about Holden's seeing the crashed spacecraft outside of Roswell, and the three small bodies of the entities who had piloted it. *See* **Holden, W. Curry**

"Scientific Study of Unidentified Flying Objects"

The study chaired by Dr. Edward Condon on behalf of the U.S. Air Force that concluded UFOs were mostly natural phenomena or

Members of the Blue Book Team. Left to right: Maj. Dewey Fournet, Jr., press liaison Al Chop, and Capt. Edward J. Ruppelt.

hoaxes, not worthy of further scientific study, and not a threat to the United States. Condon's preface to his report was the pretext for the Air Force's shutting down Project Blue Book. *See* **Condon Report; Projects, Government UFO**

Screen Memory

A psychological defense mechanism of generating false memories to protect a person from particularly painful real memories. Many psychologists have to work through these screen memories to reach the core memories that cause a patient's neurosis. UFO abduction researchers believe that aliens often implant screen memories so that abductees believe an abduction was actually something else. Abduction researchers look for various commonalties of screen memories and phobias, such as phobias of bugs or memories of small children with shiny hats or helmets, that indicate the possibility that an abduction has taken place. *See* **Abduction Scenario**

Scully, Frank

Author of *Behind the Flying Saucers*, who reported that while at a Denver University lecture given by Silas Newton in March 1950, he heard the story of a Texas oilman named Leo GeBauer, who, in turn, had heard from one of his geologists, that when the geologist had worked for the U.S. military in the late 1940s, he had been told about UFO crashes and retrievals in the American southwest.

Scully's book was one of those read by Canadian radio engineer Wilbert Smith, who used it as a basis for his interview with Robert Sarbacher. This interview has become very important to the UFO field because it became the basis for Smith's memo to the Canadian Department of Transport concerning the U.S. government's knowledge of UFOs and his recommendation that the Canadian Department of Transport initiate a project to research their existence. *See* **Sarbacher, Robert; Smith, Wilbert Brockhouse**

SDI

See **Strategic Defense Initiative**

Sea of Tranquillity

Lunar site where UFO researchers believe there exist artificial structures resembling pyramids and obelisks. The Sea of Tranquillity is also the site of the Apollo 11 landing, where the astronauts were reportedly observed by UFOs. *See* **Moon, Dark Side of the**

Search for Extraterrestrial Intelligence (SETI)

An ongoing program to receive electronic signals from extraterrestrial intelligent life. The program's vast dish arrays are tuned to receive faint radio signals from deep space and then process the signals to find patterns that may indicate intelligent design. Although SETI scientists publicly proclaim their disbelief in UFOs, they nevertheless believe that intelligent extraterrestrial life does exist. Many UFO researchers believe that should SETI actually receive a communication, it would be quickly buried by the government under the cover of national security. Indeed, many researchers believe that communications of this nature have already been received and that ongoing SETI operation is a cover for the real exchange of communications between earth and ETs. Some researchers believe that we haven't heard any ET transmissions because most civilizations keep quiet for fear that they will be found by hostile, aggressive aliens.

Senn, Charles

Former Chief of the U.S. Air Force's community relations division, who wrote in a letter to Lt. Duward Crow of NASA, "I sincerely hope that you are successful in preventing reopening of UFO investigations." *See* **National Aeronautics and Space Administration (NASA)**

SETI

See **Search for Extraterrestrial Intelligence**

Shag Harbor

A UFO incident that took place on October, 4, 1967 in Nova Scotia in which an object sixty feet long and displaying four lights descended into the water at a sharp angle. At least twenty witnesses saw the dome-shaped craft descend into the water and then drift with the ebbing tide. The object remained on the surface for about five minutes and then seemed to submerge.

Thinking that perhaps the object had actually sunk, rescuers, including local fishermen and two members of the Royal Canadian Mounted Police quickly raced to the scene to look for survivors. What they found was a half-mile long patch of thick yellow and shining foam on the surface of the water where the craft went down. Navy and police divers were brought in on October 6 to recover any wreckage, but they could not find anything.

At the time of the reported sighing there were other reports in the area of a UFO, and according to another account, a radar contact with a UFO by a fishing trawler. The Canadian government reportedly has silenced all the witnesses, both civilian and military, and the case still remains one of the unsolved mysteries of North American ufology.

Shandera, Jaime

UFO researcher who first received the MJ-12 documents in 1984. Shandera had previously worked with members of an intelligence operation involved with UFOs, who said they desired to have information about UFOs released to the public. The MJ-12 documents and the events surrounding their discovery are controversial, and have generated a debate among UFO researchers as to their legitimacy and whether researchers, themselves, are targets of a government disinformation program designed to provide them with deliberately inaccurate information. *See* **Majestic**

Shard, The

A so-called lunar anomaly, photographed by NASA's Lunar Orbiter III, a precursor to the Apollo Moon missions. The Shard appears to be a massive crystalline structure, over a mile and a half high. UFO researchers claim that it could not possibly be a naturally occurring lunar feature. Others believe it is merely an optical anomaly generated in the Orbiter's camera. *See* **Moon, Dark Side of the**

Sheehan, Daniel

Harvard Law School graduate and director of the Christic Institute, Daniel Sheehan has taught constitutional law at Antioch School of Law, the University of California at Santa Barbara, Yale Law School, Notre Dame, Cornell Law School, and Harvard. As

Chief Counsel to abduction researcher Dr. John Mack, Sheehan represented Mack before the Harvard University Faculty Committee investigating Professor Mack on the basis of his alien abduction research. Daniel Sheehan is possibly best known for his role as the chief counsel in the Karen Silkwood *v.* Kerr-McGee Nuclear Corporation. Sheehan, who has a Masters of Divinity from Harvard Divinity School, has lectured on the theological implications of the Search for Extraterrestrial Intelligence to the Jet Propulsion Laboratory.

Sign

See **Projects, Government UFO**

Sigma

See **Projects, Government UFO**

Simon, Benjamin

The psychiatrist who first treated Barney and Betty Hill for anxiety and discovered through hypnotic regression therapy that they believed they had been abducted by extraterrestrial creatures, taken aboard a spacecraft, and subjected to medical experiments. Betty Hill believes that it was someone in Dr. Simon's office who stole either the actual tape recordings of the Hills' sessions or transcripts of those recordings, and gave them to a reporter for a Boston newspaper. The reporter then published the Hills' story. *See* **Fish, Marjorie; Hill, Barney and Betty**

Sims, Derrel

UFO researcher, abductee, and abduction therapist. Sims dates his interest in UFOs from his first encounter with aliens when he was four years old. He credits his fully conscious recall of numerous abduction events with providing his first clues to the nature of the phenomenon. From 1968–71, Sims reports that he served in the U.S. Army as a senior military police officer and in a top-secret capacity with the CIA.

Derrel Sims

Sims believes that aliens are deceptive, and in pursuit of their own agenda with little regard for human casualties. When he realized that his own son was a victim of alien abduction, he became Chief of Abduction Investigations for the Houston UFO Network, and while heading the HUFON support group, claimed to initiate alien contact through the use of hypnotic suggestions inserted into a volunteer's subconscious. The purported result was a spectacular double "mass abduction" in December of 1992, involving seven individuals in two states, and offering numerous points of verification.

Sims also discovered the phenomenon of "fluorescence" on abductees. In some cases a black light shined on the skin following an abduction reveals subdermal traces, usually of brilliant green, and occasionally forming specific patterns, which Sims speculates could be a kind of alien "secretion."

Sitchin, Zecharia

Zecharia Sitchin is perhaps the foremost advocate for a human extraterrestrial heritage. A scholar of Sumerian civilization and culture, Sitchin is one of a very small number of Orientalists alive today who can read the Sumerian clay tablets of ancient Mesopotamia, which trace geological and historical events back to the dawn of civilization and beyond, into the dim past of human origins.

Zecharia Sitchin

Born in Russia and raised in Palestine, Sitchin has acquired a thoroughgoing knowledge of modern and ancient Hebrew, as well as and other Semitic and European languages, and he is a respected and long-time working scholar of the Old Testament and of the history and archaeology of the Near East. A graduate of the University of London, where he majored in Economic History, he attended the London School of Economics and Political Science, and after that became a journalist.

After his journalistic career, Sitchin began writing *The Earth*

Assyrian Cylinder Seal in which a winged, disc-shaped object can be seen on the left and a rocket-shaped object can be seen to the right of the horseman.

Chronicles, a series of books that combine textual and pictorial evidence from the past with advances of modern science to trace the story of the Anunnaki, the extraterrestrial species that Sitchin believes first came to earth some 450,000 years ago. Sitchin's cosmological theories deal with the creation of the earth through a cosmic collision with "the twelfth Planet," sometimes called "Planet X."

Zecharia Sitchin is a member of the American Association for the Advancement of Science, the American Oriental Society, the Middle East Studies Association of North America, and the Israel Exploration Society. Some of his earlier books which chronicle the saga are *The Twelfth Planet, The Stairway to Heaven, The Wars of Gods and Men, The Lost Realms, Genesis Revisited, When Time Began, and Divine Encounters*. *See* **Anunnaki**

Skunk Works

Founded by Lockheed's Kelly Johnson during the waning days of World War II, the unit got its name from its location next to a plastics factory in Burbank, California. The fumes from the factory were so intensely foul that Johnson's shop was promptly named "Skunk Works." Known for its innovative engineering and quick conception-to-operation work, the shop went on to developed some of America's most advanced aircraft, including the U-2, the SR-71, and the F-117 Stealth fighter. *See* **Johnson, Kelly; Rich, Ben**

Slayton, Donald

A Mercury astronaut, Slayton revealed in an interview that he had seen UFOs in 1951. *See* **Apollo Program**

Sleppy, Lydia

A teletype operator for radio station KSWS in Roswell, who was in the process of sending out station manager John McBoyle's "hot" news story on the Roswell crash, when she received the following teletype message: "This is the FBI. You will cease transmitting." *See* **KSWS, Roswell Radio Station; McBoyle, John; Roswell, New Mexico**

Smith, Wilbert Brockhouse

A senior radio engineer with the Broadcast and Measurements Section of the Canadian Department of Transport, Wilbert Smith made some "discreet inquiries" while in the United States concerning what the American military knew about flying saucers. He learned that a secret government study of UFOs was underway, a study as secret as the Manhattan Project during World War II. Smith sent a memo to his supervisor revealing what he had learned; in turn, the Canadian government launched its own investigation into UFOs in December of 1950, called Project Magnet, which Smith was charged with running. The top-secret memo, dated November 20, 1950, is one of the most important documents in UFO research, and reads as follows:

Area 51 in Nevada.

While in Washington attending the NARB Conference, two books were released, one titled *Behind the Flying Saucers*, by Frank Scully, and the other, *The Flying Saucers Are Real*, by Donald Keyhoe. Both books dealt mostly with the sightings of unidentified objects, and both books claim that flying objects were of extraterrestrial origin and might well be spaceships from another planet. Scully claimed that the preliminary studies of one saucer which fell into the hands of the United States government, indicated that they operated on some hitherto unknown magnetic principles. It appeared to me that our own work in geomagnetics might well be the linkage between our technology and the technology by which the saucers are designed and operated. If it is assumed that our geo-magnetic investigations are in the right direction, the theory of operation of the saucers becomes quite straightforward, with all observed features explained qualitatively and quantitatively.

I made discreet inquiries through the Canadian embassy staff in Washington, who were able to obtain for me the following information:

a. The matter is the most highly classified subject in the United States government, rating even higher than the H-Bomb.

b. Flying saucers exist.

c. Their modus operandi is unknown, but a concentrated effort is being made by a small group headed by Doctor Vannevar Bush.

d. The entire matter is considered by the United States authorities to be of tremendous significance.

I was further informed that the United States authorities are investigating along quite a number of lines which might possibly be related to the saucers, such as mental phenomena, and I gather they are not doing too well since they indicated that if Canada is doing anything at all in geo-magnetics they would welcome a discussion with suitably accredited Canadians.

Smith went on to recommend that a project be set up within his section of the Department of Transport to study the problem, and "Project Magnet" was established the following month. Although Magnet managed to confirm numerous flying saucer sightings and made that information available to the public, the Canadian government, under pressure from the United States, later adopted severe penalties for the release of UFO information, and Smith himself was forced to recant the project's findings before the Canadian House of Commons Special Committee on Broadcasting on May 17, 1955, where he said, "On the basis of our measurements, which were nil, we came to the conclusion that we had very little data of any nature to go on."

In 1950, as the basis for his original recommendation to the Canadian Department of Transport, Smith conducted an interview with Dr. Robert Sarbacher, in which he described the work he was doing and how it related to what he had been reading concerning flying saucers. Smith's notes of the interview follow:

Smith: *I am doing some work on the collapse of the earth's magnetic field as a source of energy, and I think our work may have a bearing on the flying saucers.*

Sarbacher: *What do you want to know?*

Smith: *I have read Scully's book on the saucers and would like to know how much of it is true.*

Sarbacher: *The facts reported in the book are substantially correct.*

Smith: *Then the saucers do exist?*

Sarbacher: *Yes, they exist.*

Smith: *Do they operate as Scully suggests, on magnetic principles?*

Sarbacher: *We have not been able to duplicate their performance.*

Smith: *Do they come from some other planet?*

Sarbacher: *All we know is we didn't make them, and it's pretty certain they didn't originate on earth.*

Smith: *I understand the whole topic of saucers is classified.*

Sarbacher: *Yes, it is classified two points higher than the H-bomb. In fact, its the most highly classified subject in the U.S. government at this point in time.*

Smith: *May I ask the reason for the classification?*

Sarbacher: *You may ask, but I can't tell you.*

This interview, with its admission that flying saucers exist and that the U.S. study of them is classified beyond the highest levels of secrecy, is one of the true smoking guns of UFO research. *See* **Sarbacher, Robert**

Smith, Yvonne

An abduction therapist and researcher, Smith is the founder of CERO, the Close Encounter Research Organization, a support group for individuals who have undergone alien abduction experiences. *See* **Abductees; Close Encounter Research Organization (CERO)**

Snowbird

See **Projects, Government UFO**

Snowflake, Arizona

Location where Travis Walton, subject of the motion picture *Fire in the Sky,* was working in 1975 when he was abducted by a UFO. *See* **Walton, Travis**

SOBEPS

See **Société Belge d'Étude des Phénomènes Spatiaux**

Société Belge d'Étude des Phénomènes Spatiaux (SOBEPS)

Société Belge d'Étude des Phénomènes Spatiaux (Belgian Society for Space Phenomenon Study). SOBEPS is a dossier documenting numerous UFO sightings over Belgium which suggests that UFOs are not fantasies, but real, and that they continue to make unauthorized intrusions into the secure air spaces of European nations. *See* **Belgian Triangles; OVNI, The French Report**

Socorro, New Mexico

One of a number of reported UFO crash sites in New Mexico in June and July of 1947. It is also the location where, in April 1964 Socorro police officer Lonnie Zamora discovered a landed UFO, which disembarked two strange creatures in skin-tight jumpsuits. *See* **Zamora, Lonnie**

Somebody Else Is on the Moon

George Leonard's book cataloging lunar anomalies, in which he provides photographs of "the Bridge," and large domes in the center of artificially lit moon craters. Leonard claimed that the photographs he'd found were confirmed by a NASA scientist, but that NASA had refused to disclose the information to the public. *See* **Moon, Dark Side of the**

Sotsialisticheskkaya Industriyia

A Russian scientific journal that described a Japanese expedition to the site of the 1908 Tunguska, Siberia, explosion and concluded it was not a meteorite at all but the crash of a nuclear powered spacecraft. *See* **Tunguska, Siberia**

Souers, Sidney

First director of U.S. Central Intelligence, member of the National Security Council, and a member of MJ-12. *See* **Majestic**

Space Review

The magazine published by Albert Bender in which he was set to print information about flying saucers before he was scared off by the warnings of some "Men in Black." *See* **Men in Black**

Space Telescope Science Institute (STSI)

A division of NASA that released photos of the moon from the

Hubble telescope whose resolution was so poor, UFO researchers believed they were a form of disinformation to cover up the presence of artificial structures on the moon. The Hubble photos were particularly troublesome to UFO researchers because of high resolution photos Hubble took of distant points in the galaxy. *See* **National Aeronautics and Space Administration (NASA)**

Sphinx, The

This desert enigma, standing guard over the Great Pyramid in Egypt, is thought by some to antedate Egyptian civilization by thousands of years. It has also been theorized that it may contain the records of lost civilizations such as Atlantis. Even more intriguing, duplicates of the structure are believed to exist on both Mars and Venus. The question for researchers has become: Could the Sphinx be a monument built by an earlier unknown human civilization, or one that's not of this earth?

Sphinx, Hall of Records

This is a theory that the Sphinx is not simply an Egyptian monument but antedates the Egyptian culture by thousands of years and may even be a hall of records for the Atlanteans. The work done by geologists who believe that the Sphinx suffered from water erosion and not wind erosion, while not leading necessarily to the belief that it was the final resting place of the records of the lost continent of Atlantis, nevertheless throws a new light on the possibilities that those who built the Sphinx were of a civilization possibly lost to what we take to be history, if, in fact, they are even of this earth.

Sprinkle, Leo

Counseling psychologist and author of *Soul Samples, Personal Explorations in Reincarnation and UFO Experiences,* Dr. Leo Sprinkle is also the director of the Rocky Mountain UFO Conference. Sprinkle's research has focused on the relationship among extrasensory experiences such as telepathy and manifestations of spiritual entities and UFO sightings. He has worked on a number of cases in which, he says, the contactee's perception of UFOs bore a direct relationship to the experiencer's perception of his or her mission in life.

Sputnik

Launched by Russia in 1957, this satellite was the beginning of the public space race between the United States and the Soviet Union.

SR-71 Blackbird

Designed and built by Lockheed's Skunk Works, and possibly the greatest example of "human" aeronautical engineering in the twentieth century, the Blackbird was used by the U.S. Air Force and the CIA for worldwide strategic reconnaissance. Though the Air Force would only grudgingly admit to operational capabilities "above 70,000 feet and Mach 3," those in the know believe the SR-71's real performance far exceeded those numbers. In 1991 the Air Force mothballed the Blackbird fleet, ostensibly as a cost-cutting measure, but rumors began circulating almost immediately that it had been replaced by the even more advanced "Aurora," whose technologies were claimed to come from downed UFOs. *See* **Aurora; Skunk Works**

SRI

See **Stanford Research Institute**

Stanford Research Institute (SRI)

Founded in 1946, the Menlo Park research group is best known in the UFO community for its relationship with the CIA remote viewing program. In the words of Dr. Harold Puthoff, who wrote a report documenting the CIA-SRI relationship, the remote viewing program was instituted "to determine whether such phenomena as remote viewing 'might have any utiulity for intelligence collection.' " SRI, now called SRI International, was involved with the military as the support center for the computer hosts connecting to the ARPANET, the predecessor of the Internet and was also involved in the research for the highly classified Over the Horizon Radar program. *See* **Puthoff, Hal; Remote Viewing; Swann, Ingo**

Stargate

Another name for the military's formerly top secret remote viewing program. *See* **Remote Viewing**

"Star Trek"

One of the longest-running shows in television history in its various incarnations. Many UFO researchers believe that many of the principal theories and plot lines of "Star Trek," such as the notion of the Prime Directive prohibiting interference in native cultures by more advanced ones, are not merely fictional but are actually in use today, helping to govern the protocols of contact between alien cultures and human beings.

"Star Wars"

See SDI

Stealth Aircraft

First developed at Lockheed and Northrop, these radar-evading, flying-wing-type aircraft reportedly have a shape similar to the UFO that crashed in Roswell in 1947, and may even utilize some form of electrogravitic generator culled from the Roswell saucer, rendering them even more stealthy and operationally efficient.

B-2 Stealth bomber.

Steinman, William

See Sarbacher, Robert

Stevens, Wendelle

An Air Force lieutenant colonel in the Foreign Technology Division at Wright-Patterson, who said he was "convinced beyond a doubt that we have made contact with aliens, that we are communicating with them in some way or form, and that we have vehicles and bodies in preservation."

Stevens has assembled one of the world's largest archives of UFO photographs and continues his research in photo analysis in an effort to ascertain their veracity. *See* **Air Force Foreign Technology Division**

Stock, Debbie

A MUFON investigator involved in the examination of the J. Bond Johnson photograph of Gen. Roger Ramey holding the now-famous Roswell telegram. Stock and fellow investigator Ron Regehr hired photographic analysts to evaluate the Roswell photo to see if the text of the telegram could be recovered. *See* **Ramey Telegram; Regehr, Ron**

Stone, Clifford

A former U.S. Army specialist, Stone has written that he had a UFO encounter when he was a young child in southern Ohio, an encounter that somehow got reported to the Army. After his enlistment, he was ordered to visit a special movie-screening room at Fort Belvoir, where he said he saw a film of a UFO retrieval. Stone was ultimately assigned to a Nuclear/Biological/Chemical unit—essentially a hazardous materials team—which, he says, was a cover for a UFO retrievals unit.

As a result of his experiences retrieving downed UFOs, Stone began searching for documents under the Freedom of Information Act that would allow him to disclose the government's role in covering up the existence of flying saucers and the alien entities who pilot them. His efforts have resulted in, among other disclosures, the revelation of Operations Moondust and Bluefly, dealing with the retrieval of extraterrestrial objects and the cataloging of these objects. *See* **FOIA; Operation Bluefly; Operation Moondust**

Stonehenge, England

The famous circular landmark of stones, referred to as possibly a kind of primitive solar calendar marking the winter solstice, is also the site of a large amount of crop circle actvity. *See* **Crop Circle**

Stonehill, Paul

A Russian-born journalist and writer for *UFO Magazine* in the U.S. and in the U.K. who went on to become a UFO researcher, author and lecturer. Stonehill has written articles on the Soviet espionage programs during the Cold War, parapsychology, Ukrainian history, underwater anomalies, and UFOs. His works have been widely published around the world and he has lectured extensively on the paranormal.

Strange Harvest, A

A 1980 Emmy-Award-winning documentary produced by Linda Moulton Howe that explored and attempted to understand the connection between cattle mutilations and UFOs. Howe's research led her to believe that animal mutilations come from two separate sources: UFOs, and unmarked helicopters and vans operated by branches of the U.S. government. *See* **Cattle Mutilations; Howe, Linda Moulton**

Strategic Air Command

Now defunct, this unit was commanded by Maj. Gen. Clements McMullen who ordered Col. William Blanchard to retrieve the Roswell crash debris from Chaves County Sheriff George Wilcox and send it to Fort Worth and thence to Washington, where General McMullen himself took personal possession of the material. *See* **Roswell, New Mexico**

Strieber, Whitley

Best-selling novelist and author whose most popular work, *Communion*—allegedly a novel, but based on what he later admitted were true events—was made into a motion picture by director Philippe Mora and became a seminal event in the story of UFO abductions. Strieber also authored *Majestic*, a fictional account of the Roswell crash and its coverup. *See* **Abductees**

Whitley Strieber

Stringfield, Leonard

A UFO researcher and author who cataloged many cases of UFO crashes in the American southwest and the U.S. government's retrieval of downed flying saucers and their alien pilots.

STS-48

As covered by *UFO Magazine's* research director Don Ecker and by Graham Stewart on the MUFON website, *STS-48* refers to an event that took place on September 15, 1991 between 20:30 and 20:45 GMT when a television camera mounted on the stern of the space shuttle Discovery's cargo bay that was aimed at earth's horizon captured a glowing oblong-shaped object which appeared just beneath the horizon. The object seemed to be traveling at a slow speed from the right side of the frame of the video to the left, climbing ever so slightly, when a flash emanated from the lower left of the screen and headed toward the floating object. The oblong object suddenly maneuvered into what looked like a U-turn and accelerated sharply as if to avoid the flash coming at it. Immediately after the object turned, the flash shot right through the space where the object had been. Seconds after the flash had streaked through the empty space, the camera retrained and shut down. *See* **National Aeronautics and Space Administration (NASA); Strategic Defense Initiative (SDI)**

STSI

See **Space Telescope Science Institute**

Strategic Defense Initiative (SDI)

An advanced, high-tech defense system first proposed by President Ronald Reagan in 1983, for the purpose of defending America and her allies against enemy missiles. The system was to be composed of an array of surveillance satellites to detect missile launches and track their trajectories; killer satellites to fire lasers and particle-beam weapons against the enemy missiles; earth-based laser and particle-beam weapons to destroy missile warheads that got through the satellite screen; and anti-missile missiles that would intercept the warheads that made it through the first lines of defense.

The ultimate purpose of such a system, UFO researchers believed, was to protect against a counter-attack by the Soviets in the event they mistakenly perceived a UFO sighting as a U.S. missile attack. This was not as farfetched an idea as it sounds. In a June 1989 article in *Soviet Military Review*, entitled "UFOs and Security," Aleksandr Kuzovkin states: "We believe that lack of information on the characteristics and influence of UFOs increase the threat of incorrect identification. Then mass transition of UFOs along trajectories close to that of combat missiles could be regarded by computers as an attack." *See* **Laser Weapons**

Sturrock, Peter

Professor of Applied Physics at Stanford University, winner of the 1986 Hale Prize in Solar Physics from the American Astronomical Society, the Arctowski Medal in 1990 from the National Academy of Sciences, and the 1992 Space Sciences Award from the 40,000-member American Institute of Aeronautics and Astronautics for his "major contribution to the fields of geophysics, solar physics and astrophysics, leadership in the space science community, and dedication to the pursuit of knowledge."

Peter Sturrock

Peter Sturrock chaired the Rockefeller Panel at Pocantico Hills, New York, and oversaw the writing of what has been called "The Sturrock Report," examining those aspects of the UFO phenomenon that cannot be dismissed as either hoaxes or natural occurrences and which, therefore, require further scientific evaluation. Even without plaudits from the scientific community at large, the value of this special panel report is unmistakable.

In his introduction to the report, Professor Sturrock emphasized that he saw the work of his panel as a type of response to the Condon Report, suggesting those areas where science needs to explore in order to unravel some of the mysteries posed by the UFO enigma. *See* **Condon Report**

Sturrock Report, Rockefeller Panel

Author Scott Smith, writing in *UFO Magazine*, suggests that the most important event in the history of studying the UFO phenomenon was the release of a report from a panel of independent scientists sponsored by the Society for Scientific Exploration on June 29, 1998, also called "The Sturrock Report." The conclusion of this panel, after considering evidence for the reality of UFOs presented by investigators, was: "[T]here was no convincing evidence pointing to the involvement of extraterrestrial intelligence. The panel nevertheless concluded it would be valuable to carefully evaluate UFO reports since, whenever there are unexplained observations, there is the possibility that scientists will learn something new by studying these observations."

The co-chairs of the panel were Dr. Thomas Holzer of the High Altitude Observatory of the National Center for Atmospheric Research in Boulder, Colorado, and Dr. Von Eshleman, Emeritus Professor of Electrical Engineering at Stanford. Other panel members included Dr. J. R. Jokipii and Dr. H. J. Melosh, professors of planetary science at the University of Arizona in Tucson; Dr. James Papike, director of the Institute of Meteoritics and Professor of Earth and Planetary Sciences at the University of New Mexico in Albuquerque; Dr. Charles Tolbert, Professor of Astronomy at the University of Virginia; Dr. Francois Louange, Managing Director of Fleximage in Paris; Dr. Guenther Reitz of the Germany Aerospace Center, Institute for Aerospace Medicine in Cologne, Germany; and Dr. Bernard Veyret of the Bioelectromagnetics Laboratory at the University of Bordeaux, France. Serving as the panel moderators were two scientists with previous interest and involvement in cutting-edge science: David Pritchard and Harold Puthoff.

The report, which was published with a commentary by Peter Sturrock in 1999, documents celebrated cases of UFO encounters such as the Mansfield, Ohio Army helicopter case and the Cash-Landrum case and explains why these events require further scientific study since they cannot be explained away by conventional science. *See* **Cash-Landrum Case; Mansfield, Ohio, Army Helicopter Case**

Swann, Ingo

Artist, author, and gifted psychic, Ingo Swann is one of the best known veterans of the CIA remote viewing program, which program he began, according to Dr. Harold Puthoff of Stanford Research Institute, after sending Puthoff a brief proposal about the potential benefits of a government-funded psychic research program. Swann is also credited with naming the process *remote viewing*.

Ingo Swann

Swann believes that all human beings are endowed with different types of intuition and telepathic awareness, which allow them to retrieve information by viewing it psychically. Human beings are also able to transfer that information telepathically, Swann has written, and thus has been central to the development of the government's remote viewing operations. *See* **Remote Viewing; Stanford Research Institute (SRI)**

Szolnok, Hungary

A site in Hungary from which many reports of UFO sightings had been received, according to George Keleti, the Hungarian Minister of Defense.

TASS

The Russian national news service that historically presented the Soviet government's view of the news. After the collapse of the Soviet Union, *TASS* was more open about reporting events and has been forthcoming about reporting UFO sightings and encounters in Russia. *See* **Russian Roswell**

"Tercer Milenia" ("Third Millennium")

Mexican television news magazine show anchored by Jaime Maussan, one of the leading commentators on UFO events. "Tercer Milenia" has broadcast incredible footage of UFO sightings and showcased some of Mexico's most bizarre cases of UFO encounters, even as American television has shied away from serious coverage of similar types of UFO stories. *See* **Maussan, Jaime**

Tesla, Nikola

One of the most influential of early twentieth-century scientists/ inventors. Tesla discovered alternating current and an efficient means for transmitting it over long distances and was the scientist to whom a federal court in 1943 awarded the title *Inventor of Radio.* Tesla advocated the concept of "free energy," the transmission of particle-beam energy from huge towers to power distribution nodes over long distances; a completely wireless form of transmission. He also believed that he was capable of communicating with extraterrestrials through ultra-high-powered radio beams.

Nikola Tesla

Tesla spent years actively trying to sell his advanced weapons ideas to the U.S. Navy and Army, but was consistently rebuffed by the Naval Research Board run by Thomas Edison before World War I and again by the War Department during World War II. In fact, buried in Gen. Nathan Twining's files at the Air Materiel Command were portions of Tesla's notes on directed particle beams that Office of Alien Property agents, under the direction of the FBI, seized from Tesla's New York City hotel room the day after he died—files that have remained "buried" to this day.

A dark, brooding Serbian, Tesla could never understand why George Westinghouse, J. P. Morgan, John Astor, and other millionaire financiers could not, or would not, see the benefits of Tesla's bolts of pure electrical energy beamed from one point on earth to another. What Tesla himself couldn't see was that neither Morgan nor Westinghouse was interested in energy they couldn't charge customers for.

With the modern military's growing interest in particle-beam weaponry and power transmission, it appears that Tesla's inventions, once given short shrift, will be what future generations of technology are based upon.

Tower, The

An immense lunar structure that has been photographed by astronauts and earlier moon probes. The Tower is five miles high, and stands close to "the Shard," another crystalline-appearing structure. The Tower is comprised of cubes, some as large as a mile wide. *See* **Moon, Dark Side of the**

Traders, Alien Beings

According to contactees, the Traders are androgynous beings who have traveled the universe for billions of years and long ago discovered a way to channel energy from black holes to supply their planet. Contactees have reported that the Traders discovered a way to "lasso" small black holes and draw them close to their planet, "feeding" the black holes to sustain the energy balance. Unfortunately, the black holes grew larger the more they consumed, and began to pull the Traders' home world into them. The Traders fled their planet, an event that so devastated them that they took on the mission to preserve all life forms in the universe by protecting the means to sustain life.

Tranquillity Base

First lunar base established by Apollo 11 after their module landed on the moon. *See* **Apollo Program; Moon, Dark Side of the**

Travelers, Alien Beings

A variety of alien races who travel along with the Traders to different worlds. According to contactees, there are currently at least nineteen different races that have achieved Traveler status. Several of the races are directly involved with monitoring humanity and tracking our growth as a species. Contactees have said that we have nothing to fear from the Traveler races because any Traveler who causes harm to another species loses its right to travel with the Traders. It is the Traders, reportedly, who govern the Travelers and impose upon developing species, like ours, limits to their ability to travel in space so as to avoid interfering with or waging war on other races. Among the Traveler races are:

Artist's conception of a "Mantis" being, based on abductee reports.

The Zetas

Zetas are the small grays most commonly identified as aliens or ETs, and come from the star system Zeta Reticuli. Zetas are the creatures the U.S. Army recovered from the crashed spaceship at Roswell. Like the Traders, the Zetas do not communicate via spoken language but instead transmit their thoughts telepathically.

The Nordics

Tall, blond, and almost indistinguishable from human beings, this race of beings, contactees say, act as caretakers, and their members are quite benevolent. Because of the Nordics' human-like appearance, many contactees don't even realize that they are of extraterrestrial origin and believe them to be angelic creatures. The Nordics have told their contactees that they are here to observe us. They find human beings intriguing because of our individuality and our refusal to recognize limitations to what we can accomplish.

The Mantis Beings

To their human contactees, the Mantis Beings appear as six-foot-tall creatures that resemble a praying mantis. They have a specific role in the contact/abduction experience, paralyzing their subjects with an electrical pulse to prevent them from injuring themselves because of the human fright/flight reaction. In this capacity, the Mantis Beings act almost like anesthesiologists during the contact experience.

The Reptilians

Another six-foot-tall race, these creatures, contactees have said, have actually contributed DNA strands to our genetic makeup. These beings are explorers, mapping out distant parts of the universe for their Trader supervisors. The Reptilians have taught their contactees that the entire planet Earth is alive and that our distinctions between animal, plant, and mineral are based upon completely false assumptions. In fact, they have said, every object on our planet is alive and to a degree sentient in its own way. Contactees have explained that the Reptilians are the beings carrying the message that humanity is destroying earth's resources. Accordingly, they have taught their contactees to be messengers of preservation.

Large Grays

Unlike the Zetas, large grays seem to have a supervisory rather than a drone-like role. Betty Hill remembers that one larger creature in the group that abducted her and her husband Barney seemed to behave as if he were an officer, taking on the primary role of interrogator and directing the smaller grays to perform their respective tasks.

Travis Air Force Base

A strategically important Air Force base in the United States where, in 1960, all its bombers were put on red alert for a counterstrike against the Soviet Union when a formation of targets was detected by air defense radars flying via a North Pole route towards the United States. Suddenly the targets disappeared from radar, the bomber crews were ordered to stand down, and the mistaken alert was blamed on radar reflections from the moon.

Trindade Island

See **Brazilian Navy Case**

Truman, Harry S.

The plain-speaking former U.S. senator from Missouri was Franklin Roosevelt's vice president during Roosevelt's fourth term, and upon FDR's death in 1945 became the thirty-third president of the United States. It was Truman's decision to drop the world's first atomic bomb on Japan, as well as (so the story goes) deal with the crash of a flying saucer outside of Roswell, New Mexico, in 1947. Truman created the super-secret government committee MJ-12 to handle the UFO "problem."

President Truman's role in MJ-12 is set forth in the controversial Majestic documents, defining the mission as investigative as well as administrative. The documents include a comprehensive briefing for incoming president Dwight D. Eisenhower on what the military found at Roswell and what the implications of the discovery are.

Despite controversy over the nature of the MJ-12 briefing documents and the President's alleged role in the whole affair, Truman's statement at an April 4, 1950, press conference, in response to a question on flying saucers, is a matter of public record: "I can assure you that flying saucers, given that they exist, are not constructed by any power on earth." *See* **Majestic**

Tujunga Canyon Abductions

A California abduction case from the early 1950s in which two women were startled by lights outside their remote rural canyon home. Their experience came long before the Betty and Barney Hill, Copley Woods, and Brooklyn Bridge abductions, and before the story of the Roswell crash became public—so the women had no idea what an alien abduction was. Although they were fearful about the lights in such a remote location, they suddenly realized that hours had passed—time they couldn't account for. Subsequent therapy revealed that the two women had been abducted and subjected to medical examinations by strange creatures.

Tunguska, Siberia

The site of a tremendous explosion on June 30, 1908. The explosion was so great that its fireball was seen by people five hundred miles away. The equivalent of a thousand Hiroshima atomic bombs going off at once, the blast flattened an entire forest of trees for hundreds of miles in all directions.

There are a number of theories to explain the explosion, the most mainstream of which is that a large meteor or comet entered earth's atmosphere and exploded over the area. However, Japanese scientists who visited the region were convinced the explosion and devastation resulted from the uncontrolled chain reaction of a nuclear reactor on board a spacecraft. The Japanese expedition even marked the site with a plaque commemorating it as an extraterrestrial event. Author Linda Moulton Howe has claimed that she was told by an official inside the U.S. intelligence community that they could confirm the cause of the disaster was an exploding spacecraft—because our alien "guests" had told them so. *See* **Podkamennaya Tunguska River; *Sotsialisticheskkaya Industriyia***

Twining, Nathan

Commanding officer of the Air Materiel Command based at Wright-Patterson Air Force Base in 1947, who played a pivotal role in the official military investigation into the crash at Roswell. Shortly after the crash, Twining canceled a preplanned trip and made an emergency visit to New Mexico. The initiation of a cover story, substitution of Roswell debris with that of a weather balloon, and subsequent USAAF pronouncements denying a flying saucer had been found, all followed Twining's sudden visit.

Gen. Twining later became Chairman of the Joint Chiefs of Staff, Chief of Staff of the U.S. Air Force, and, reportedly, a member of MJ-12. In a September 23, 1947, memo to the commanding general of the Air Force, Twining wrote: "The phenomenon reported is something real and not visionary or fictitious ... There are objects probably approximating the shape of a disc, of such appreciable size as to appear to be as large as a man-made aircraft." Twining's memo is considered to be one of the initiating factors in the establishment of MJ-12. *See* **Majestic**

Nathan Twining

Twinkle, Project

An investigation conducted principally by Dr. Lincoln La Paz of the University of New Mexico that involved the study of fiery balls of light—"green meteors"—that appeared over New Mexico in the 1940s and, according to some people, still appear today. Whether a form of ball lightning, some other natural occurrence—or an entirely unknown phenomenon—the study was cut short when the project, originally scheduled to be conducted by the Air Force, never received its funding, for reasons still unclear. Rudimentary attempts to record the balls of light on film were unsuccessful, and the phenomenon has remained an unsolved mystery.

U-2 Spy Plane

America's first extreme-high-altitude surveillance jet aircraft, used by the CIA/U.S. Air Force for strategic reconnaissance over Soviet Bloc and Eastern Bloc nations. Designed and built by Lockheed's Skunk Works in 1955 and first test-flown at Area 51 in Nevada, it was a U-2 piloted by Francis Gary Powers that was shot down by the Soviet Union over Russia in 1960. The CIA's Haines Report suggested that the spy agency used UFO sightings as cover for their U-2 flights. *See* **Central Intelligence Agency, History of UFOs in**

UFO Controversy in America, The

Doctoral dissertation and later a book published by David Jacobs in 1975. Jacobs' study of UFOs as an aspect of popular culture was perhaps the first Ph.D. dissertation on the subject of UFOs.

UFO

See **Unidentified Flying Object**

UFO Healings

A book by Preston Dennett that documents over a hundred cases of abductees who have been cured from grave illness by aliens using medical technology more advanced than our own.

UFO Reporting

Although the common theory is that most people who report UFOs are somehow delusional, deceived by their own eyes, or outright hoaxers, UFO researchers are quick to point out that at least two U.S. presidents have gone public with their own UFO sightings: Jimmy Carter, by filing an official report, and Ronald Reagan, in an interview with the *Wall Street Journal*.

A UFO sighting is best witnessed by multiple observers, if possible, because each can confirm the general details of the sighting. Witnesses should try to jot down as many of the details surrounding the event as possible, including, but not limited to, the direction of object(s), size and speed of the object(s), presence of any other airplanes or helicopters in the area, time of day, and nature of the weather. Sightings of UFOs should be reported to MUFON, which can be located on the Internet at www.mufon.com, where visitors will find phone numbers, fax numbers, and e-mails to make confidential reports.

UFO Incident, The

NBC Movie of the Week dramatizing the abductions of Barney and Betty Hill and starring James Earl Jones and Estelle Parsons. *See* **Hill, Barney and Betty**

UFO Magazine

The science and phenomena magazine founded in 1986 by Vicki Cooper and Sherie Stark; it is the only magazine continuously published in the United States dedicated to publishing articles about UFOs, alien encounters, and related scientific phenomena.

Just about every prominent researcher in the field as been interviewed by members of the magazine staff, no matter how controversial; thus readers are able to make up their own minds as to various claims

UFO Magazine is the longest continually running publication devoted to the phenomenon of alien contact in the world.

about topics as far-ranging and varied as alien abductions, government conspiracies, military mind-control experiments, visits by strange beings in ancient times, as well as coverage of UFO reports around the world. *See* **Ecker, Don; Ecker, Vicki**

UFO Photographs Around the World, Vol. 1

A book by Col. Wendelle Stevens, reviewing his photo-investigative procedure, and containing a narrative of experiences witnesses have had upon encountering objects from another world. *See* **Stevens, Wendelle**

"UFOs and Security"

The lead article in the June 1989 issue of *Soviet Military Review*, written by Aleksandr Kuzovkin, stating that UFOs could be deemed a threat to a nation's security simply because of their flight patterns and the inability of a country's defense commanders to distinguish them from flights of enemy missiles, leading to a mistaken nuclear counterstrike. *See* **Kuzovkin, Aleksandr; North Pole; Strategic Defense Initiative (SDI)**

Unidentified object captured on Russian radar. This image was one of literally thousands researchers have discovered after the collapse of the Soviet Union and the release of Russian UFO files.

UFOs: Interplanetary Visitors

A book written by Raymond Fowler and J. Allen Hynek in which engineer Paul Santorini suggests that one of the reasons the existence of UFOs has been covered up is that governments of the world will not disclose to their respective populations that there is a force against which even the most powerful countries cannot mount a defense. Dr. Santorini said he had been consulted by a number of scientists concerning the existence of UFOs. *See* **Santorini, Paul**

"UFOs Through the Eyes of Cosmonauts"

An article in which Soyuz-5 cosmonaut Yevgeni Khrunov states that the aerodynamic attributes of flying saucers, specifically their speed and ability to maneuver, "simply stagger the imagination."

UFO Wave Over Belgium: An Extraordinary Dossier

An extensive record of observations of UFOs over Belgium, both civilian and military, confirming sightings of craft displaying behavior and characteristics unlike any known earth craft. *See* **SOBEPS**

Uhouse, Bill

Witness to a Kingman, Arizona, flying saucer landing in 1953. Uhouse claims that aliens landed their craft and gave it to the military so as to lessen the military's need to shoot them down.

Uhouse said that the saucer was left on the ground with the hatch open. The first soldiers to enter it became disoriented and nauseous. The disk was transported to what became Area 51, where it kept making loud humming noises that frightened the workers, forcing the military to leave the craft out on the tarmac. Uhouse has also claimed that he worked alongside gray aliens while at Area 51. *See* **Area 51; Oz Effect**

Underwater UFOs

UFOs are said to be not only capable of navigating through earth's atmosphere, but also of traveling great distances under our oceans. There are numerous reports of underwater UFO bases, such as off Puerto Rico (The Bermuda Triangle), and in the Eastern Pacific, and of encounters with these craft either traveling through, entering, or exiting oceans and lakes.

Filiberto Cardenas, in a 1979 encounter in Florida, claimed that he was abducted by a UFO and taken aboard while his wife and daughter looked on. Cardenas was kept aboard for two hours while the craft dove into the ocean and headed for the aliens' underwater base. *See* **Cardenas, Filiberto**

"Unidentified Flying Objects"

In the entry in the *Encyclopedia Americana,* astronomer Carl Sagan suggests that it seems clear to scientists that the earth is not the only inhabited planet in the known universe since evidence points to the existence of other planetary systems around stars. Sagan speculates that there are perhaps a million planets in our own galaxy that may host life forms that are technologically far superior to us.

At its most basic, a UFO is exactly what its acronym represents: a flying object which cannot be identified by its observer and whose definition still defies identification after it has been studied by others. Once the object is identified, whether as a plane, flying saucer, or natural phenomenon, then it is no longer unidentified.

A flying saucer is not technically a UFO once it has been identified. Though the overwhelming majority of lay people and reporters refer to flying saucers as UFOs (and to UFOs as flying saucers), this is logically incorrect.

According to the U.S. Air Force, a UFO is any "aerial object which the observer is unable to identify." The 1969 Condon Report from the University of Colorado described a UFO as:

Viking test craft that landed at White Sands in 1972.

> The stimulus for a report made by one or more individuals of something seen in the sky (or an object thought to be capable of flight but seen when landed on earth) which the observer could not identify as having an ordinary natural origin, and which seemed to him sufficiently puzzling that he undertook to make a report of it to police, to government officials, to the press, or perhaps to a representative of a private organization devoted to the study of such objects.

According to astronomer and UFO researcher J. Allen Hynek, UFOs are "The reported perception of an object or light seen in the sky or upon the land, the appearance, trajectory, and general dynamic and luminescent behavior of which is only mystifying to the original percipients but remains unidentified after close scrutiny of all available evidence by persons who are technically capable of making a commonsense identification, if one is possible." *See* **Condon Report**

United Nations

Although not ordinarily known for its UFO investigations, the UN General Assembly and Security Council have been a forum for UFO and ET pronouncements. In 1978, U.S. astronaut Gordon Cooper appeared before the General Assembly to ask for "open discussions" on unidentified spacecraft. Almost a decade later, President Ronald Reagan challenged UN members to lay aside their differences if Earth were faced with a threat from outer space. *See* **Gairy, Eric**

U.S. Army Research and Development

The office where Lt. Col. Philip Corso spent his final years in the Army and where, under the command of Lt. Gen. Arthur Trudeau, Corso "seeded" the artifacts from the Roswell file into those industries already working on similar technology.

Phillip J. Corso, left, and Lt. Gen. Arthur Trudeau, right, commander of US Army Research and Development.

Corso's memoirs about his years in Army Research and Development are published in his book *The Day After Roswell*. *See* **Corso, Philip**

U.S. National Archives

The national repository where many UFO-related documents may still reside, even though they may not be filed or readily accessible.

U.S. Office of War Information

Essentially a propaganda ministry during World War II, the Office of War Information published an information manual for the motion picture industry in 1943 in which it urged filmmakers and studios to support the war effort affirmatively by "making" each civilian understand "that he is an integral part of the war effort, and if he loses the war, he loses everything." This example of information molding via the media continued through the 1940s and beyond as the government policy to deride those who claimed UFO encounters, and punish those in government, the military, and in influential positions within the scientific community who dared breach the government's order of silence regarding UFO issues.

U.S. Research and Development Board

A high-level panel of bureaucrats and scientists in charge of shaping government/military policies in regard to advances in technology. Dr. Robert Sarbacher was a consultant to the board, where he worked with others who reportedly had connections with MJ-12. *See* **Majestic; Sarbacher, Robert**

"Unsolved Mysteries"

American television show which has covered many flying saucer and UFO-related stories, including the RAF Bentwaters case and crop circle sightings.

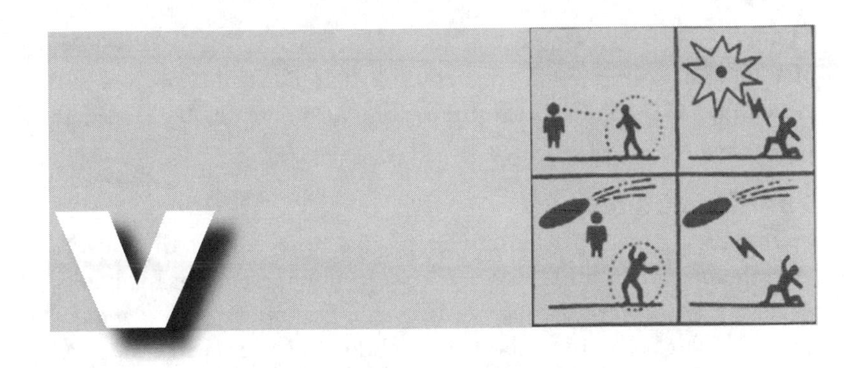

"V," The Series

One of the more celebrated television series dramatizing the colonization of planet Earth by extraterrestrials who, though they appear to be friendly and offer advanced science and technological gifts to humanity, actually seek to harvest human beings and plunder earth's resources. The aliens co-opt earth governments into keeping their citizens complacent about the alien presence. Though they appear to look like humans, these aliens are actually horrific lizard-like creatures with an appetite for human flesh. Could the lizard creatures of "V" actually have inspired UFO witnesses who claim to have seen Reptilians? *See* **Reptilians**

V-2 Rocket

A German ballistic missile, carrying an explosive warhead, launched at England during World War II. The United States captured a number of V-2s at war's end, along with the German rocket scientists who developed them, and brought them all back to test facilities in America. Many believed that it was a V-2 rocket, launched from the White Sands (New Mexico) missile-test range, that had crashed outside of Roswell in 1947. Records confirmed, however, that no V-2s were launched during the few days when the wreckage of the spacecraft was discovered.

Valenov Plate

A thousand-year-old plate found in the mountains of Nepal upon which is depicted a flying saucer and an alien being. The plate is made of a combination of metal and ceramic, sculpted from a process that even today is difficult, if not impossible, to duplicate. *See* **Däniken, Erich von**

Vallee, Jacques

The French astrophysicist who has become one of the world's leading authorities on UFOs and related phenomena. In his books, including *Fastwalkers, Passport to Magonia, Dimensions,* and *Forbidden Science,* Vallee has written about alien abduction, the science supporting the possibilities of UFOs, and the relationship among various types of psychic phenomena, folklore, and alternative or extradimensional universes.

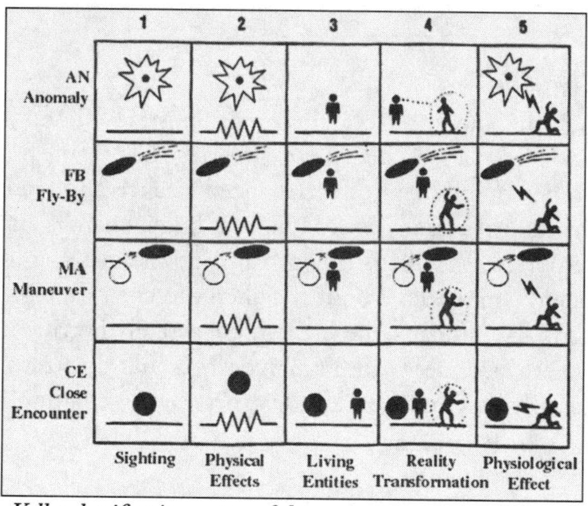

Vallee classification system of close encounters with UFOs and extraterrestrials, which shows the effects UFO encounters have on human beings.

He is perhaps best known for expanding upon the original Hynek classification system, which Dr. Hynek invented to categorize levels of UFO encounters. In particular, the system organizes what are called "close encounters" into encounters that range from visual sightings within five hundred feet to encounters of the fifth kind during which communication between a human and extraterrestrial takes place. *See* **Brookings Institution; Hynek, Allen.**

Vandenberg Air Force Base

U.S. Air Force base near Lompoc, California, which is an important missile-launch facility and a site where numerous flying saucer sightings have been reported and landings may have taken place in the early 1950s.

Vandenberg, Hoyt

Former chief of Staff of the U.S. Air Force, and former director of central intelligence before the CIA came into being. Gen. Vandenberg reportedly was in charge of security for MJ-12. *See* **Majestic**

Van Flandern, Tom

The former chief of celestial mechanics at the U.S. Naval Observatory and a Yale Ph.D. in astronomy. Dr. Van Flandern told the 191st American Astronomical Society conference that the Face on Mars was not a optical illusion or a naturally occurring feature, but something possibly made by intelligent entities. Basing his controversial conclusions on his new analysis of the existing data received from Mars probes, Dr. Van Flandern said, "I suggest that in view of these test results, we prepare ourselves for a cultural shock certainly unrivaled in modern times." *See* **Mars, Face on**

Varginha, Brazil

Site of a UFO encounter in January of 1996. Extraterrestrials were not only observed by local residents, but they interacted with several of the citizens and were photographed during the encounter. Dr. Roger Leir, the podiatric surgeon from California who has performed numerous extractions of alien implants, has researched the Varginha case, interviewed witnesses to the encounters, and claims to have in his possession a body part from one of the aliens. *See* **Leir, Roger**

Vatican

The "headquarters" of the Roman Catholic Church maintains an ongoing intelligence network to investigate and report on paranormal events around the world. In part, the Vatican needs to determine whether incidences of miracles, apparitions of saints or other figures, and supernatural events have any substance to them. The Holy See also maintains the Papal Observatory in the United States and is actively involved in investigations into the existence of UFOs.

Intelligence community sources both within the United States and in Europe report that the Vatican has long known about the existence of extraterrestrials and may even have a direct downlink from the Hubble Space Telescope for its own personal viewing. The Vatican's insider on matters supernatural and extraterrestrial is Monsignor Corrado Balducci, also known as the "Pope's Exorcist," who has openly discussed alien encounters and admitted that he is a member of a Vatican council whose responsibility it is to figure out a way to handle the existence of extraterrestrials. *See* **Balducci, Corrado**

Velikovsky, Immanuel

Immanuel Velikovsky

Freudian psychiatrist who developed a theory of ancient catastrophes, suggesting they were caused by a natural, albeit cosmological extraterrestrial event—the near collision between the earth and the planet Venus. Velikovsky's book *Worlds in Collision*, which was condemned by the formal academic community in 1951, nevertheless became a bestseller. Velikovsky was driven out of academia and blacklisted within the American academic community, as were his supporters.

One of Velikovsky's most vehement critics was astronomer Dr. Donald Menzel, who was also a vicious debunker of UFOs while at the same time listed as a member of the top-secret UFO study group MJ-12. As UFO researcher and author Stanton Friedman has suggested, Menzel's public stance as astronomer and debunker would have been perfect cover for a member of MJ-12.

In his book, Velikovsky explored the correlations between the Biblical plagues and natural catastrophes, which he suggested were caused by extraterrestrial events. He noted in a passage in the "Book of Joshua," that a shower of meteorites fell on the valley of Beth-horon in the time of Joshua. In addition, he learned that during the long day of Joshua, Mexican annals stated that "the world was deprived of light and the Sun did not appear for a fourfold night." In the Caribbean, "the Sun rose only a little way over the horizon, and remained there without moving; the Moon also stood still." The correlation of the long day of Joshua in Israel with the long night in the Americas made astronomical sense to Velikovsky. In fact, years after *Worlds in Collision* was published, Sumerian civilization and "Tenth Planet" scholar Zecharia Sitchin was able to synchronize these events based on historical documents from various cultures.

Velikovsky began to research the myths and legends of ancient peoples from across the globe and correlated human records with what the ancients described as an immense cataclysm associated with an object seen in the sky. Although the established scientific and academic community dismissed these accounts as unverifiable, Velikovsky persisted and developed a theory that there was indeed just such a connection between these catastrophes and the planet Venus.

From one ancient civilization to another, Velikovsky noted that the people associated their respective catastrophes with the planet Venus, describing the planet as "comet-like." The Mexicans, for example, called Venus a "star that smoked," their phrase to describe a comet. In the Hindu Vedas, Venus is depicted with "fire accompanied by smoke." "Fire is hanging down from the planet Venus," states the Hebrew Talmud. In Egypt, Venus as Sekhmet was "a circling star which scatters its flames of fire." The Aztecs called Venus *Quetzalcoatl,* or "plumed serpent," whose feathers are acknowledged to signify "flames of fire."

The dragon or serpent is one of the most universal glyphs for the comet in the ancient world. Thus, the pillar of light in the sky by day, by which the Israelites traveled, and the pillar of fire by night, by which the Israelites encamped, as described in the "Book of Exodus," Velikovsky concluded, was the planet Venus moving in an elliptical orbit that brought it close enough to Earth to create a global upheaval.

Based on evidence he developed from these ancient histories, Velikovsky argued that perhaps around 10,000 years ago the planet Jupiter became unstable when a great mass of solar-system debris fell into it, causing it to spin rapidly. This instability in turn caused Jupiter to fission off part of its stony-iron core, as well as thousands of cometary bodies, moving in countless orbits throughout the solar system. The largest of these bodies headed toward the Sun and Earth, coalescing into the planet Venus. Velikovsky derived this process from the English cosmologist Raymond Lyttleton, who claimed that this was how the other terrestrial-type planets such as Mercury and Mars may have been born.

For an indeterminate time, Venus moved on its elongated path around the sun, nearly colliding with the earth around 1,500 BC, during the time of Exodus. Earth's axis was changed, bringing wholesale destruction to civilization and life. Fifty years later, Venus caused a second catastrophe in the time of Joshua, but since its orbit had changed and it was now more distant from the earth, the devastation was less dramatic.

Velikovsky was roundly attacked by rank-and-file astronomers. However, Albert Einstein suggested that he could find nothing wrong with Velikovsky's hypothesis. Nevertheless, the censorship persisted. Harlow Shapley was a Harvard University astronomer, and in league with several of his colleagues attempted to prevent Velikovsky's publisher, Macmillan, from going forward with publishing the book. When this failed, they organized a boycott of Macmillan's textbooks by universities, which brought the company to its knees. To salvage this lucrative part of its business, Macmillan transferred its rights to *Worlds in Collision* to Doubleday, which didn't have a textbook subsidiary that could be blackmailed. Macmillan was then forced to burn the rest of the copies in its possession and fire editor James Putnam, who'd handled the book for the company.

A decade later, Alfred De Grazia, editor of the *American Behavioral Scientist,* with science historian Livio Stecchini and engineer Ralph E. Juergens, wrote *The Velikovsky Affair*, which exposed the entire matter and pointed out that Velikovsky's critics were themselves guilty of intellectual dishonesty and coercion. The book, while a bombshell, ultimately changed few minds in the academic and scientific communities. Today, while Velikovsky's theories are well regarded, the man himself is still deemed a "crackpot" by many. *See* **Majestic; Menzel, Donald; Sitchin, Zecharia; Venus**

Venus

The enigmatic, cloud-shrouded planet on whose surface, some scientists say, NASA has discovered an array of pyramids and a Sphinx-like structure, not unlike the structures that are thought to exist on Mars. Planetary theorist Immanuel Velikovsky claimed Venus had nearly collided with earth in biblical times and was responsible for many of the catastrophic events documented in the Book of Exodus. *See* **Velikovsky, Immanuel**

Vietnam War, UFO Appearances

According to retired Army Specialist and UFO researcher Clifford Stone, there were numerous UFO encounters during the Vietnam War, including an incident in which both American and NVA/VC forces had agreed to a brief cease-fire so as to allow each side to remove its dead and wounded from a no-man's-land in a free-fire zone. A UFO appeared overhead, according to Stone, surveilling the battlefield while both sides simply watched in awe. *See* **Stone, Clifford**

Viking

The 1976 NASA Mars probe whose photographs revealed the presence of a Sphinx-like humanoid face on the Martian surface. *See* **Mars, Face on**

Vorilhon, Claude

See **Rael**

Voronezh, Russia

In 1989 several children from the city of Voronezh, three hundred miles south of Moscow, saw a strange craft land nearby from which emerged two large beings and a small "mechanical man." According to *TASS*, the Soviet/Russian news agency, the children were playing at the city park in the early evening when they witnessed what has been variously described as a "pink, shining" object in the sky and as a "red-colored ball that measured around thirty feet in diameter." The children, as well as an assembling crowd, "could clearly see a hatch opening in the lower part of the ball and a humanoid in the opening. A three-to-four-meter-tall, small-headed, three-eyed being—wearing silvery overalls and bronze boots—was seen inside the sphere."

The creature emerged alongside another similar being and "a robot." A shining triangle, measuring thirty by fifty centimeters, according to witnesses, appeared on the ground and then vanished. The robot reportedly moved in a mechanical fashion when one of the aliens touched it. A howl of fear came from a sixteen-year-old boy, who was silenced and paralyzed by an alien, whose eyes were "shining." One of the creatures carried a long tubular "gun" by his side. He pointed the device at the teenager, who vanished. When the aliens later left in their "ball," the boy reappeared.

TASS later reported more UFO sightings in Voronezh during the same time period, with dozens of witnesses claiming to have found UFO landing "traces" such as circular indents in the soil, as well as strange footprints. There was widespread skepticism generated by the *TASS* reports on the "Russian UFO Invasion," with questions regarding the apparent lack of reaction from the Soviet Defense Ministry.

Voyager 1 Spacecraft

A NASA planetary probe launched in 1977 to explore the outer planets, destined to fly beyond the limits of our solar system. Voyager s equipped with messages for any species the probe may encounter.

Wackenhut Corporation

A privately owned and run security firm whose directors have close ties to the United States military and its intelligence services. Wackenhut provides perimeter security for Area 51 as well as for many other sensitive government and military installations.

Wallace, Mike

One of the hosts of the popular CBS television news show "60 Minutes," and a pioneer of television news broadcasting. In 1958 Wallace hosted Maj. Donald Keyhoe for what turned out to be quite a contentious television interview.

Keyhoe, who often criticized the U.S. Air Force for their public dismissal of UFO reports, asserted that the hundreds of professional observers, both military and civilian, who had seen UFOs could not all be hallucinating and suggested that the Air Force itself in the early 1950s had believed that UFOs were real, but by the end of the decade was denying its own beliefs to the American people. Keyhoe went on to cite a litany of evidence he believed proved the existence of flying saucers.

Wallace, in response, claimed that most Americans believed that flying saucers were not real, but more likely the product of "pranksters." A skeptical Wallace continued to probe Keyhoe's views on the subject, and ended up taking him to task for criticizing the Air Force. While news shows of the period were usually circumspect in their reporting of events, Wallace's curious lack of objectivity about UFOs was alarming, and it can now be seen, in retrospect, as a foretaste of today's more slanted media coverage. *See* **Keyhoe, Donald**

Walters, Ed

The controversial central character in the Gulf Breeze, Florida-sightings case, and author of a book by the same name. Walters told the story of how one evening he had spotted a flying saucer directly over his house and that it had paralyzed him with a blue light. It then attempted to "beam him up" while he fought to free himself. It was the beginning of Walters' long struggle against a group of aliens that continued to track his every move, communicated with him telepathically, and tried to lure him out into the open again so as to abduct him.

Walters took a number of Polaroid snapshots of the UFOs which were authenticated by some photographic experts, while others discount them; he received the support of MUFON International Director Walt Andrus. Walters was later accused of hoaxing the incident, and his credibility took some major hits. In his defense it must be noted that his actions were responsible for drawing into the open many other witnesses in the community, all of whom confirmed that there was a major UFO flap taking place in the skies over Gulf Breeze. *See* **Andrus, Walt; Gulf Breeze Incident**

Walters, Frances

Ed Walters' wife and also a participant in the Gulf Breeze sightings who responded to attacks on her husband by saying that the sightings were real events whose impact changed the lives of her family and the Gulf Breeze community as well. *See* **Gulf Breeze Incident; Walters, Ed**

Walton, Travis

The self-described abductee whose story was recounted in the motion picture *Fire in the Sky.* In 1975, while clearing brush for the National Forest Service just outside of Snowflake, Arizona, Walton was abducted by a UFO as his fellow crew members looked on helplessly. After driving back to town and reporting their friend's disappearance, the local sheriff wanted to charge the men with murder. But their story repeatedly passed lie detector tests, and before the sheriff could file any charges, Walton himself reappeared. He had been "missing" for six days.

Walton never wavered in his description of the abduction experience and of meeting small gray aliens and taller Nordics. He continues to lecture on it today. While the incident was bizarre, Walton's friends have continually backed him up and have never changed their stories.

War of the Worlds, The Radio Broadcast

Originally a novel by English science-fiction writer H. G. Wells, describing an attack upon planet Earth by invaders from Mars, the story was adapted for broadcast on American radio by Howard Koch and broadcast by theatrical entrepreneur and soon-to-become famed director Orson Welles.

On Sunday night, Halloween night, October 31, 1938, Welles and his Mercury Theatre Group presented what has become the single most famous performance in the history of American broadcasting. Welles presented *The War of the Worlds* as if it were an actual news broadcast covering the progress of an alien invasion, all of which began in what was then a picturesque little rural hamlet just outside of Princeton, New Jersey, called Grovers Mills.

For the next hour, Welles, as an increasingly frantic newscaster, described the inexorable advance of the Martian invaders towards New York, overcoming local police units and then pushing right through the U.S. Army itself with their unstoppable siege machines. As other Martian ships landed, disgorging their war machines, it looked hopeless for the beleaguered military, whose hour of ultimate defeat was drawing closer.

The radio program was repeatedly interrupted by station breaks, along with disclaimers that what was being broadcast was a dramatization. However, for those who had tuned in between announcements, the sheer authenticity of the presentation led many listeners to believe a real invasion from Mars was underway. As the dramatized Martian attack pressed on towards New York, the sense of general panic among the gullible grew, and many people reportedly prepared to flee for their lives.

If the reports of panic in the streets were true, it was unfortunate that so many missed the ending of the broadcast. Because in the end, it was the Martians who ultimately lost the "war"—defeated not by standing armies of withering firepower or by the indomitable will of earthmen, but by the luckiest of chances that the Martians had no defense against the common earthly cold virus. During and at the conclusion of the broadcast, the radio station's telephone switchboard was swamped with distraught, and very angry, callers.

Though in later years it's been suspected that the reports of widespread panic and some deaths were exaggerated by the media, UFO researchers believe the public's response, limited as it may have been, led military and government officials to the decision that the American people must never be told the truth about UFOs in order to avoid chaos. Nine years after the broadcast of *The War of the Worlds,* in a not-quite-so-picturesque little desert hamlet called Roswell, that decision took full effect. *See* **Brookings Institution**

Warren, Larry

U.S. Air Force enlisted man posted to RAF Bentwaters who was part of a security detail on December 26, 1980, that investigated a curious light that appeared to fall into Rendlesham Forest near the perimeter of the base. Warren later wrote in his book *Left at Eastgate* (co-authored with Peter Robbins) that he saw what he thought was a spacecraft with strange creatures inside. Warren also claims to have witnessed an encounter between the base deputy commander and the creatures inside the craft. *See* **Bentwaters Case**

Washington, D.C., UFO Overflights

One of the most controversial sightings in UFO history. On the evenings of July 19 and 20, 1952, eight flying disks were observed traveling directly over the White House, the Capitol building, and the Pentagon. The objects were tracked visually from the ground, by air-defense radar from the control towers at D.C.'s National Airport and Andrews Air Force Base, and by interceptor pilots who had been scrambled to drive the intruding craft out of restricted airspace.

The saucers cruised slowly at first, then accelerated to astounding velocities as they passed over the Capitol building. Jet interceptors gave chase, but to no avail, as the saucers outran, out-maneuvered, and simply outflew anything the military defense command put in the air. The saucers suddenly disappeared from radar screens, then reappeared miles away. This was the continuing pattern for two successive nights, before the intrusions halted altogether. However, a week later, on July 26 and 27, the saucers reappeared, and again were able to consistently penetrate restricted airspace. The "UFOs over the Capitol" story made headlines in all the local newspapers.

Finally, USAF Maj. Gen. John Samford called a press confer-

UFOs over Washington DC. Here lights fly over the United States Capitol Building.

ence in an effort to squelch the furor over the "saucer invasion." The military's explanation was that the sightings were illusions caused by temperature inversions that created "ghost" images in the air and on radar. However, the pilots who chased the flying disks later told UFO investigators that the objects were real, but they had been ordered by their superiors to forget what they saw and deny that they had made contacts with any targets. To date there has never been an adequate explanation for the UFOs seen over Washington in 1952. *See* **Bolling Air Force Base; Samford, John**

Washington Post

One of the few major newspapers to feature with any prominence the photos of obelisks on the moon taken by a lunar orbiter. On November, 22, 1966, the *Post* headline read: "Six Mysterious statuesque Shadows Photographed on the Moon by Orbiter." *See* **Moon, Dark Side of the**

Weiner, Jack and Jim

Twin brothers who had been abducted during the "Allagash Affair." After the abduction, Jack found a lump on his leg. When a doctor examined it he found a strange-looking object had been implanted beneath Jack's skin. The doctor sent the object to the Centers for Disease Control in Atlanta, but the Air Force seized control of it and refused to release any information about it. Jack has been a multiple abductee; he and his wife were abducted years after Allagash. Their full story is related in Ray Fowler's *The Allagash Abductions.*

Jim Weiner had been suffering from a form of epilepsy in the limbic region of his brain, known as a "waking dream" state. Asked by his doctor about the types of dreams he experienced, Jim explained that they usually consisted of him awaking from sleep to discover strange creatures standing at the foot of his bed. Jim's doctor referred him to UFO researcher Raymond Fowler, who began regression treatment and uncovered the story of Jim's and Jack's abductions on the Allagash Waterway. Jim also learned through a series of regression-therapy sessions that he and Jack had been frequent abductees since they were babies by a group of aliens particularly fascinated by twins. *See* **Allagash Affair; Leir, Roger**

Weiner, Mary

Jack Weiner's wife who was abducted along with him a few years after the Allagash Affair. *See* **Allagash Affair**

Welles, Orson

One of America's most important film directors. Welles' Mercury Theatre Group radio play of H. G. Wells' *The War of the Worlds* caused an East Coast, as well as nationwide, furor in 1938—"free" publicity which, some insist, Welles was not adverse to.

After the broadcast, Welles was forced by New York City officials and broadcast network executives to issue a public apology. While mustering all his considerable talent, Welles was able to convey both remorse and impish pleasure in his statement. At a time when real war clouds were gathering over Europe, the firestorm generated by a fictional Martian invasion was too much irony to be lost on the actor-director. *See War of the Worlds, The* **Radio Broadcast**

Wen-Qwang

In the *Journal of UFO Research,* the People's Republic of China, in 1981, Wen wrote, "China is so vast, and UFOs are certainly being witnessed again and again all throughout China."

Wheless, Hewitt T.

Assistant Vice Chief of Staff of the U.S. Air Force. In March of 1967, Lt. Gen. Wheless sent out a memo to various agencies within the Department of Defense reporting that persons in civilian clothes and claiming to be representatives of the Defense Department were contacting civilians who had UFO encounters, warning them to cease talking about their experiences. In the memo, Lt. Gen. Wheless specifically disavowed any DoD connection with these unidentified individuals and asked all DoD employees to report any knowledge of them to the Pentagon's internal security office. *See* **Men in Black**

White, Ed

Gemini astronaut who, with James McDivitt in June, 1965, saw a strange-looking metallic object with long arms protruding from it as they were orbiting over the Hawaiian Islands. *See* **McDivitt, James**

White Sands Proving Grounds

This highly secure and important facility in New Mexico used for testing American guided missiles was overflown by flying saucers on June 29, 1947, and again on July 2, just days before the crash outside of nearby Roswell. According to a memo from the Navy Department to the Director of the CIA, the proving grounds at White Sands had been continually buzzed by flying saucers for years after 1947. White Sands was the launch site for the V-2 rockets captured from the Germans at the end of World War II, one of which was cited as the real source of the crash debris at Roswell. *See* **V-2 Rocket**

Whitmore, Walt, Jr.

Son of the owner of Roswell radio station KGFL, who was also a witness to the Roswell crash debris. Whitmore described the mysterious metallic fabric and "I-beams" retrieved from the wreckage as "very much like tinfoil in appearance, but could not be torn or cut at all, extremely light in weight. Some small beams that appeared to be either wood or wood-like had a sort of writing on it which looked like numbers which had either been added or multiplied." *See* **Roswell, New Mexico**

Wilkins, H. P.

An English astronomer who, in 1953, confirmed the sighting made by *New York Herald Tribune* science editor John J. O'Neill of a twelve-mile long bridge on the surface of the moon. Wilkins told the BBC that "It looks artificial. It's almost incredible that such a thing could have formed in the first instance, or if was formed, could have lasted during the ages in which the moon has been in existence." *See* **O'Neill, John; Moore, Patrick**

Wilcox, George

George Wilcox

Sheriff of Chaves County, New Mexico, who took delivery of a box of debris from Mac Brazel a day after Brazel discovered the Roswell crash debris field on the Foster Ranch. After seeing the debris, Sheriff Wilcox thought it strange enough to call the Roswell Army Air Field, where he spoke to Maj. Jesse Marcel, the base intelligence officer.

Sheriff Wilcox put the box of debris in a jail cell for safekeeping, but it was retrieved very quickly thereafter on orders from Col. William Blanchard, the commander of the 509th Bomber Squadron headquartered at the base. Shortly after his call to the air base, Sheriff Wilcox was paid a visit by military police who warned him never to discuss the incident, or he and his family would be killed.

Wilcox's close friend and fellow witness, Glenn Dennis, had also been threatened into silence. As he tells the story, he and Wilcox were frightened by the whole affair. So whenever they wanted to vent their feelings about the Army's cavalier treatment and the threat that was hanging over their heads, they would find a spot in the desert where they knew no one could eavesdrop, and talk about it. *See* **Children of Roswell**

Wilcox, Ines

George Wilcox's wife has become one of the unsung heroes of Roswell because of a diary she kept during her husband's term of office as Chaves County sheriff. In it, she recorded how Mac Brazel brought a box of strange debris to the county jail for safekeeping and how the Army showed up the next day, took the debris, and swore her husband, George, to silence. *See* **Children of Roswell**

Woodbridge Royal Air Force Base

The neighboring base to RAF Bentwaters. Both bases were involved in the appearance of what Larry Warren and others called a UFO on the nights of December 26 and 27, 1980, when the base deputy commander, according to witnesses, actually communicated with extraterrestrials in a spacecraft that had landed in Rendlesham Forest just outside the base perimeter. *See* **Bentwaters Case**

Wolfe, Karl

Karl Wolfe

A writer for *UFO Magazine* and a former consultant to UPN's KCOP in Los Angeles, Wolfe reported an experience in which he viewed first-hand classified U.S. Air Force photos of artificial structures on the moon.

Stationed at Langley Field, Virginia in 1965 with the 4444 Reconnaissance Technical Group at Headquarters Tactical Air Command, Wolfe worked for the director of intelligence at the field, which was the hub for the Lunar Orbiter project. Wolfe was only one of two technicians authorized to work with the classified photos taken by the U-2 spy planes and surveillance satellites. On a day when Wolfe was the only technician on duty at the base, he was asked to repair a special type of photo-imaging and enhancing device used by NASA, which was occupying their own separate facilities on a different part of the base.

The photo-enhancing printer Wolfe was asked to repair was located in a darkroom where he would be unable to run the type of diagnostics that would enable him to locate the problem in the device. While he waited for a crew to remove the device to a secure location where he could work on it, the other airman in the darkroom, assuming that Wolfe had the requisite security classification to view the types of photos the enhancing device printed, showed Wolfe photos of what looked like a group of artificial structures on the dark side of the moon.

The airman explained that NASA had discovered a "base" on the dark side of the moon, a base that didn't belong to the U.S. government nor to any government on Earth. Wolfe said that one of the structures seemed to have an array of circular antennas and prompted him to think that the base was a joint venture between the military and extraterrestrials. *See* **Moon, Dark Side of the**

Woolsey, James

Former director of Central Intelligence, who in 1997 was asked about UFOs and responded by commissioning a report for the open CIA publication *Studies in Intelligence*, titled "CIA's Role in the Study of UFOs, 1947-1990." *See* **Central Intelligence Agency, History of UFOs in**

Wright-Patterson Air Force Base

Located in Dayton, Ohio, Wright-Patterson was the home of the Air Materiel Command under the leadership of Lt. Gen. Nathan Twining, later a member of MJ-12. The alien bodies and crash debris from the UFO crash outside of Roswell were taken to Wright-Patterson in the days following the retrieval, where it remained stored into the 1960s, according to Arizona Senator Barry Goldwater. *See* **Goldwater, Barry; Hanger 18; Majestic**

X-15

An experimental high-altitude, high-speed aircraft flown out of Edwards Air Force Base in California in the 1950s and 1960s. X-15 test pilots recorded numerous UFO encounters during this period. On one such mission in July of 1962, Maj. Robert White, during a 58-mile-high flight, observed a UFO tracking his aircraft, and radioed to his controllers on the ground, "I have no idea what it could be. It's grayish in color and about thirty to forty feet away. There are things out there! There absolutely is!"

"X-Files," The

An anthemic television show that ran on the Fox Network during the 1990s, which depicted a determined, almost obsessive FBI agent and his skeptical partner on the hunt for the truth—which he knew was out there—about UFOs, alien abductions, and the band of complicit government and business officials who were not only covering up the truth but conducting a private war with the aliens.

Pulling incidents from bits of UFO lore, the "X-Files" garnered a loyal audience of fans, some of whom believed that the show was the government's way of filtering the truth to the American people. It was the first television show in decades, and one of the most successful in television history, to channel the paranoia about an alien presence and alien abduction conspiracies to a receptive, mass-market television audience.

X-Rays

In their first incarnation they were "shadowgraphs" invented by Nikola Tesla and were the basis for Roentgen in his later discovery of X-rays. These particles can change into rays and then back into particles. X-rays have become among the most important tools in modern diagnostic medicine and have been used to discover the devices implanted into human abductees by the extraterrestrials who abducted them and who continue to monitor their whereabouts through these telemetric subcutaneous implants. *See* **Implants, Alien; Tesla, Nikola**

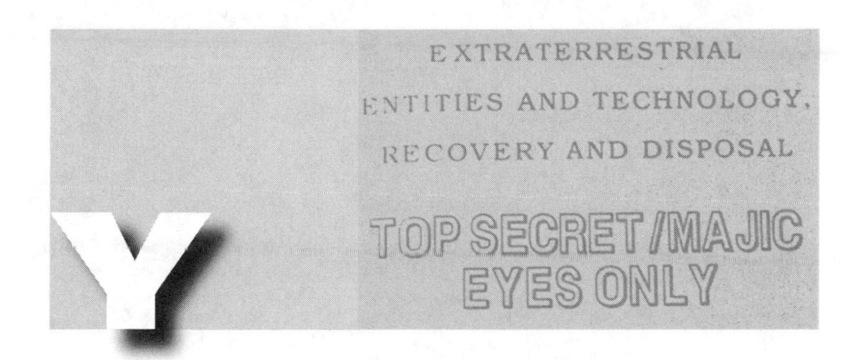

Yale University

Final resting place of the bulk of the KGB files from the Soviet era, which detail many of the reported UFO encounters by Soviet pilots and military personnel, and the KGB investigations into their veracity. *See* **KGB Files**

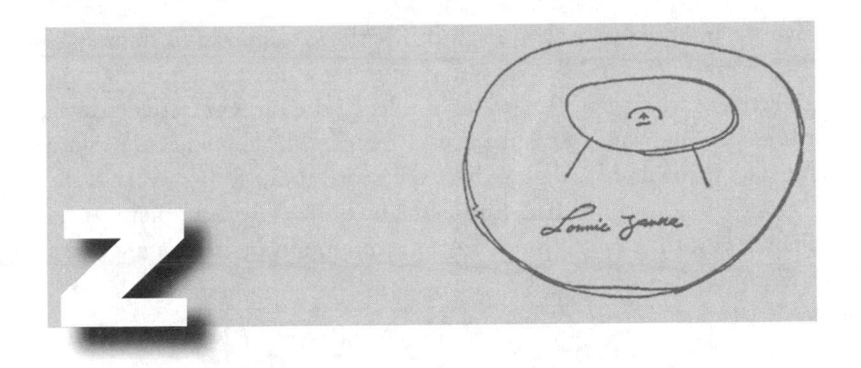

Zacatecas Observatory

A Mexican observatory where the first UFO photos were taken in 1883. Professor A.Y. Bonilla captured the images of 116 disks moving across the face of the Sun.

Zamora, Lonnie

One of the best-documented cases of a law enforcement UFO sighting. On April 24, 1964, in Socorro, New Mexico, Officer Lonnie Zamora was chasing a speeder into the outskirts of town at around 5:45 in the evening when he saw a flaming object shoot by him and heard what sounded like an explosion. He terminated his pursuit to investigate what he thought might be the detonation of a dynamite shack nearby. He followed a dirt path that led to the shack and saw that it was intact, but in the sky above was a smokeless blue and orange sliver of flame descending to the desert floor.

As Zamora drove closer, he noticed the flame and roar had stopped. About two hundred yards away, he saw a shiny object that he later said resembled "a car turned upside down … standing on its radiator or trunk." A closer look revealed "two people in white coveralls. One of these persons seemed to turn and look straight at my car and seemed startled—seemed to quickly jump somewhat."

Zamora went on to describe the occupants as "normal in shape, but possibly they were small adults or large kids." After radioing for backup, Zamora quietly walked to within about a hundred feet of the object. He later described the UFO to FBI Special Agent J. Arthur

Byrnes, Jr. as being oval and smooth with no windows or doors visible. It sat on girder-like legs, and Zamora saw a red insignia that was about two-and-a-half feet wide. Suddenly, the object emitted a low-frequency roar that rose until it was "very loud." The object let out a burst of flame and kicked up dust, then rose about fifteen feet off the ground. It cleared the dynamite shack by only a few feet. Silently and flamelessly, the object climbed over a mountain in the distance and was gone.

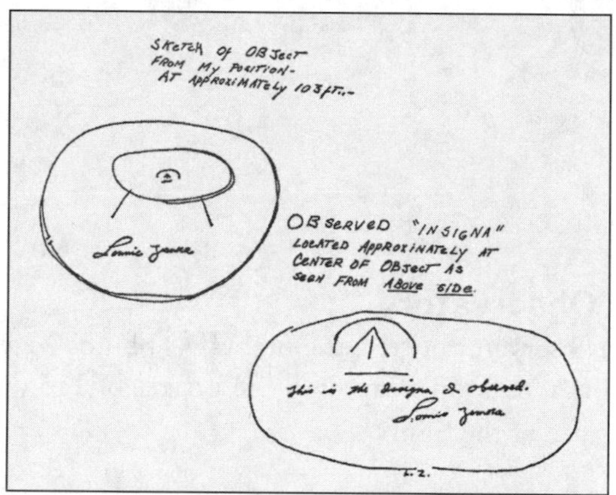

Lonnie Zamora's sketch of the UFOs he saw in Socorro, New Mexico.

A few minutes later, Sgt. Sam Chavez of the New Mexico State Police arrived. He did not see the craft, but did witness the still-smoldering brush where it had touched down. Special Agent Byrnes, Chavez, and Deputy Sheriff James Lucky examined the area. They found four burn marks and four V-shaped depressions that were between one and two inches deep and eighteen inches long.

Zamora believed the "legs" had left these imprints. Later, an engineer's analysis of the area revealed that for the marks to be pressed so deeply into the desert, each of the "legs" had borne a load of at least one ton.

UFO debunker Philip Klass suggested Officer Zamora made up the account to increase tourism in the area. At least the author of one article, in the Central Intelligence Agency's publication, *Studies in Intelligence*, did not agree. In the article, Hector Quintanilla, then head of the Air Force's Project Blue Book, stated: "There is no doubt that Lonnie Zamora saw an object which left quite an impression on him.

There is also no question about Zamora's reliability. He is a serious officer, a pillar of his church, and a man well-versed in recognizing airborne vehicles in this area. He was puzzled by what he saw, and frankly so are we. This is the best documented case on record." *See* **Police Sightings of UFOs; Quintanilla, Hector**

Zavodsk Square, Voronezh, Russia

A park where in 1989, a UFO landed and startled a group of children. When one of the children charged the group of aliens that had emerged from the craft, the alien aimed a device at the boy, who then disappeared. He reappeared after the craft had taken off. This sighting was so incredible that *TASS* carried the story and it was even reported in the *New York Times* on October 9, 1989. *See* **Voronezh**

Zechel, Todd

One of the founders of Ground Saucer Watch and co-founder, along with Peter Gersten, of CAUS, Citizens Against UFO Secrecy. CAUS has been instrumental in filing lawsuits against the federal government for covering up information about UFOs. *See* **Gersten, Peter**

Zero-Point Propulsion

A method of propulsion that relies on the conversion of gravity, a reharnessing of a force that exists everywhere in the known universe and is, therefore, inexhaustible and recyclable.

UFO researcher Paul Hill credits Professor Hermann Oberth as being the first scientist to conclude that "UFOs convert gravitational-field energy for propulsive purposes." Zero Point theory suggests that empty space is really not empty at all, but rather a reservoir of gravitic energy that can be turned on and off, according to Dr. Hal Puthoff, by "quantum fluctuations." If this reservoir of endless power could be processed by some sort of engine, it would be theoretically capable of generating enormous speeds that could literally bend the space-time continuum so as to enable travelers to journey through the medium of time as well as through space. *See* **Oberth, Hermann**

Zetas

The "little gray" aliens that most abductees report having seen. These creatures are about four feet tall, hairless, with large black, almond-shaped eyes, and have tiny slits for mouths. Some researchers believe them to be bioengineered creatures, androids capable of traversing vast distances in space. Zetas come from the star system of Zeta Reticuli, communicate with their abductees telepathically, and seem to be capable of some sort of DNA transfer process that is gradually either hybridizing our species or implanting our DNA in other races in the galaxy. In this respect, they are spreading our species throughout the galaxy to insure that life will flourish on other planets. *See* **Alien Beings; Grays**

Zeta Reticuli

The twin-star system Zeta 1 and Zeta 2, thirty-seven light years distant from earth, whose suns are a billion years older than our sun. It was Betty Hill who first drew the star map that she said the Zetas showed her, a map that pinpointed the Zetas' home world, and whose authenticity and accuracy was later confirmed by the release of the *Gliese Catalog of Stars. See* **Fish, Marjorie; Hill, Barney and Betty;** ***Gliese Catalog***

Zigel, Felix Y.

In a November 10, 1867 Soviet television broadcast, Dr. Zigel of the Soviet Scientific Commission said:

> Unidentified flying objects are a very serious subject which we must study fully. We appeal to all viewers to send us details of strange flying craft seen over the territories of the Soviet Union. This is a serious challenge to science and we need the help of all Soviet citizens.

And less than three months later, Dr. Zigel was quoted in an article in *Soviet Life* as saying:

> Observations show that UFOs behave sensibly. In a group-formation flight, they maintain a pattern. They are most often spotted over airfields, atomic stations, and other very new engineering installations. On encountering aircraft, they always maneuver so as to avoid direct contact. A considerable list of these seemingly intelligent actions gives the impression that UFOs are investigating, perhaps even reconnoitering. The important thing now is for us to discard any preconceived notions about UFOs and to organize on a global scale a calm, sensation-free and strictly scientific study of this strange phenomenon. The subject and aims of the investigation are so serious that they justify all efforts. It goes without saying that international cooperation is vital.

A

B

M

T

About the Editors

William J. Birnes, Ph.D., publisher of *UFO Magazine,* is the New York Times bestselling author, with Lt. Col. Philip J. Corso, of *The Day After Roswell* (Pocket Books, 1997) and the editor-in-chief of the *McGraw-Hill Personal Computer Programming Encyclopedia* (1985, 1989). He is also the author of the current *Riverman* with Robert Keppel (Pocket, revised edition 2004) and also with Robert Keppel, the *Psychology of Serial Killer Investigations* (Elsevier, 2003) and the forthcoming *A Worker in the Light* with "Coast to Coast AM" host George Noory (Tor/Forge 2004). A National Endowment for the Humanities Fellow, Dr. Birnes is married to Nancy Hayfield and lives in New York and Los Angeles.

Nancy Hayfield Birnes is a novelist, nonfiction writer, editor, and president of Filament Books, an ebook publishing company in New York and Los Angeles. She has written *Cleaning House* (Farrar Straus & Giroux, 1980), *Cheaper and Better* (HarperCollins, 1988), *ZapCraft* (Ten Speed, 1990), and is also the president of the book production company Shadow Lawn Press. A 1979 *summa cum laude* graduate of Princeton, Nancy Hayfield taught writing at Trenton State College.

Vicki Cooper Ecker is the editor-in-chief and founder of *UFO Magazine.* She is currently writing a book about her years in the fields of the strange and the paranormal.

Don Ecker, a member of an Army security special operations unit in Vietnam and a medically retired police detective, is the research director of *UFO Magazine.* Ecker is a novelist and is currently the host of the radio talk show "Dark Matters."

Associate editor **Harold Burt** is a director of the Orange County chapter of MUFON and the author of *Flying Saucers 101.*

Ron Press is the managing editor of *UFO Magazine.*